MITOCHONDRIAL BIOLOGY: NEW PERSPECTIVES

The Novartis Foundation is an international scientific and educational charity (UK Registered Charity No. 313574). Known until September 1997 as the Ciba Foundation, it was established in 1947 by the CIBA company of Basle, which merged with Sandoz in 1996, to form Novartis. The Foundation operates independently in London under English trust law. It was formally opened on 22 June 1949.

The Foundation promotes the study and general knowledge of science and in particular encourages international co-operation in scientific research. To this end, it organizes internationally acclaimed meetings (typically eight symposia and allied open meetings and 15–20 discussion meetings each year) and publishes eight books per year featuring the presented papers and discussions from the symposia. Although primarily an operational rather than a grant-making foundation, it awards bursaries to young scientists to attend the symposia and afterwards work with one of the other participants.

The Foundation's headquarters at 41 Portland Place, London W1B 1BN, provide library facilities, open to graduates in science and allied disciplines. Media relations are fostered by regular press conferences and by articles prepared by the Foundation's Science Writer in Residence. The Foundation offers accommodation and meeting facilities to visiting scientists and their societies.

Information on all Foundation activities can be found at http://www.novartisfound.org.uk

Novartis Foundation Symposium 287

MITOCHONDRIAL BIOLOGY: NEW PERSPECTIVES

John Wiley & Sons, Ltd

Copyright © Novartis Foundation 2007
Published in 2007 by John Wiley & Sons Ltd,
　　　　　　　　The Atrium, Southern Gate,
　　　　　　　　Chichester PO19 8SQ, UK

　　　　　　　　National　　　01243 779777
　　　　　　　　International (+44) 1243 779777
　　　　　　　　e-mail (for orders and customer service enquiries): cs-books@wiley.co.uk
　　　　　　　　Visit our Home Page on http://www.wileyeurope.com
　　　　　　　　or http://www.wiley.com

All Rights Reserved. No part of this book may be reproduced, stored in a retrieval system or transmitted in any form or by any means, electronic, mechanical, photocopying, recording, scanning or otherwise, except under the terms of the Copyright, Designs and Patents Act 1988 or under the terms of a licence issued by the Copyright Licensing Agency Ltd, 90 Tottenham Court Road, London W1T 4LP, UK, without the permission in writing of the Publisher. Requests to the Publisher should be addressed to the Permissions Department, John Wiley & Sons Ltd, The Atrium, Southern Gate, Chichester, West Sussex PO19 8SQ, England, or emailed to permreq@wiley.co.uk, or faxed to (+44) 1243 770620.

This publication is designed to provide accurate and authoritative information in regard to the subject matter covered. It is sold on understanding that the Publisher is not engaged in rendering professional services. If professional advice or other expert assistance is required, the services of a competent professional should be sought.

Other Wiley Editorial Offices

John Wiley & Sons Inc., 111 River Street, Hoboken, NJ 07030, USA

Jossey-Bass, 989 Market Street, San Francisco, CA 94103-1741, USA

Wiley-VCH Verlag GmbH, Boschstr. 12, D-69469 Weinheim, Germany

John Wiley & Sons Australia Ltd, 33 Park Road, Milton, Queensland 4064, Australia

John Wiley & Sons (Asia) Pte Ltd, 2 Clementi Loop #02-01, Jin Xing Distripark, Singapore 129809

John Wiley & Sons Canada Ltd, 6045 Freemont Blvd, Mississauga, Ontario, Canada L5R 4J3

Wiley also publishes its books in a variety of electronic formats. Some content that appears in print may not be available in electronic books.

Novartis Foundation Symposium 287
x + 244 pages, 32 figures, 2 tables

Anniversary Logo Design: Richard J Pacifico

British Library Cataloguing in Publication Data

A catalogue record for this book is available from the British Library

ISBN 978-0-470-06657-7

Typeset in 10½ on 12½ pt Garamond by SNP Best-set Typesetter Ltd., Hong Kong
Printed and bound in Great Britain by T. J. International Ltd, Padstow, Cornwall.
This book is printed on acid-free paper responsibly manufactured from sustainable forestry, in which at least two trees are planted for each one used for paper production.

Contents

Symposium on New perspectives on mitochondrial biology, held at the Novartis Foundation, London, 28–30 November 2006

Editors: Derek J. Chadwick (Organizer) and Jamie Goode

This symposium is based on a proposal by Michael Duchen

David G. Nicholls Chair's introduction 1

Albert Neutzner, Richard J. Youle and **Mariusz Karbowski**
Outer mitochondrial membrane protein degradation by the proteasome 4
Discussion 14

Sarah E. Haigh, Gilad Twig, Anthony A. J. Molina, Jakob D. Wikstrom, Motti Deutsch and **Orian S. Shirihai** PA-GFP: a window into the subcellular adventures of the individual mitochondrion 21
Discussion 36

Luca Scorrano Multiple functions of mitochondria-shaping proteins 47
Discussion 55

Bruce M. Spiegelman Transcriptional control of mitochondrial energy metabolism through the PGC1 coactivators 60
Discussion 63

Charles Affourtit, Paul G. Crichton, Nadeene Parker and **Martin D. Brand**
Novel uncoupling proteins 70
Discussion 80

Cecilia Giulivi Mitochondria as generators and targets of nitric oxide 92
Discussion 100

György Hajnóczky, Masao Saotome, György Csordás, David Weaver and **Muqing Yi** Calcium signalling and mitochondrial motility 105
Discussion 117

Anna Romagnoli, Paola Aguiari, Diego De Stefani, Sara Leo, Saverio Marchi, Alessandro Rimessi, Erika Zecchini, Paolo Pinton and **Rosario Rizzuto** Endoplasmic reticulum/mitochondria calcium cross-talk 122
Discussion 131

Brian O'Rourke, Sonia Cortassa, Fadi Akar and **Miguel Aon** Mitochondrial ion channels in cardiac function and dysfunction 140
Discussion 152

Paolo Bernardi and **Michael Forte** The mitochondrial permeability transition pore 157
Discussion 164

Dominic James, Philippe A. Parone, Olivier Terradillos, Safa Lucken-Ardjomande, Sylvie Montessuit and **Jean-Claude Martinou** Mechanisms of mitochondrial outer membrane permeabilization 170
Discussion 176

M. Flint Beal Mitochondria and neurodegeneration 183
Discussion 192

Mügen Terzioglu and **Nils-Göran Larsson** Mitochondrial dysfunction in mammalian ageing 197
Discussion 208

Eric A. Schon and **Salvatore DiMauro** Mitochondrial mutations: genotype to phenotype 214
Discussion 226

Contributor Index 234

Subject index 236

Participants

Vera Adam-Vizi Department of Medical Biochemistry, Semmelweis University, PO Box 262, H-1444 Budapest, Hungary

M. Flint Beal Department of Neurology and Neuroscience, Weill Medical College of Cornell University, 525 East 68th Street, Room F610, New York, NY 10021, USA

Piotr Bednarczyk (*Novartis Foundation Bursar*) Department of Biophysics, Agricultural University SGGW, 159 Nowousynowska St., 02-776 Warsaw, Poland

Paolo Bernardi Department of Biomedical Sciences, Università di Padova, Viale Giuseppe Colombo 3, I-35121 Padova, Italy

Martin D. Brand MRC Dunn Human Nutrition Unit, Hills Road, Cambridge CB2 2XY, UK

Michael Duchen Department of Physiology, University College London, Gower Street, London, WC1E 6BT, UK

Cecilia Giulivi Department of Molecular Biosciences, University of California, 1311 Haring Hall, One Shields Avenue, Davis, CA 95616, USA

György Hajnóczky Department of Pathology, Anatomy and Cell Biology, Thomas Jefferson University, Rm 253 7AH, 1020 Locust Street, Philadelphia, PA 19107, USA

Andrew P. Halestrap Department of Biochemistry, School of Medical Sciences, University of Bristol, Bristol BS8 1TD, UK

Derek Hausenloy (*Novartis Foundation Bursar*) The Hatter Cardiovascular Institute, University College Hospital, 67 Chenies Mews, London WC1E 6HX, UK

Howard T. Jacobs Institute of Medical Technology, University of Tampere, FI 33014, Finland

Nils-Göran Larsson Karolinska Institutet, Department of Laboratory Medicine, Division of Metabolic Diseases, Novum, S-14186 Stockholm, Sweden

John J. Lemasters Pharmaceutical Sciences and Biochemistry & Molecular Biology, Medical University of South Carolina, QF308 Quadrangle Building, 280 Calhoun Street, PO Box 250140, Charleston, SC 29425, USA

Jean-Claude Martinou Department of Cell Biology, Sciences III, University of Geneva, 30 quai Ernest Ansermet, 1211 Geneva 4, Switzerland

David G. Nicholls (*Chair*) Morphology Core, Buck Institute for Age Research, 8001 Redwood Blvd, Novato, CA 94945, USA

Brian O'Rourke School of Medicine, Institute of Molecular Cardiobiology, The Johns Hopkins University, 720 Rutland Avenue, 1059 Ross Bldg., Baltimore, MD 21205-2195, USA

Sten Orrenius Institute of Environmental Medicine, Division of Toxicology, Karolinska Institutet, Box 210, Stockholm, SE-171 77, Sweden

Anant Parekh Department of Physiology, Anatomy and Genetics, University of Oxford, Parks Road, Oxford, OX1 3PY, UK

Ian J. Reynolds Merck Research Laboratories, WP42-229, 770 Sumneytown Pike, P O Box 4, West Point, PA 19486-0004, USA

Peter R. Rich The Glynn Laboratory of Bioenergetics, Department of Biology, University College London, Gower Street, London WC1E 6BT, UK

Rosario Rizzuto Department of Experimental and Diagnostic Medicine, General Pathology Section, University of Ferrara, Via L. Borsari 46, 44100 Ferrara, Italy

Eric A. Schon Department of Neurology, Room 4-431, College of Physicians and Surgeons, Columbia University, 630 West 168th Street, New York, NY 10032, USA

Luca Scorrano Dulbecco-Telethon Institute, Venetian Institute of Molecular Medicine, Via Orus 2, Padova, I-35129, Italy

PARTICIPANTS

Orian Shirihai Department of Pharmacology and Experimental Therapeutics, Tufts University School of Medicine, 136 Harrison Avenue, Boston, MA 02111, USA

Bruce M. Spiegelman Dana-Farber Cancer Institute and Department of Cell Biology, Harvard Medical School, Boston, MA 02115, USA

Douglas M. Turnbull Mitochondrial Research Group, School of Neurology, Neurobiology and Psychiatry, The Medical School, Newcastle University, Newcastle Upon Tyne, NE2 4HH, UK

Richard J. Youle Biochemistry Section, Surgical Neurology Branch, National Institutes of Health, Bldg 35, Room 2C917, 25 Convent Drive, MSC 370, Bethesda, MD 20892, USA

Chair's introduction

David G. Nicholls

Buck Institute for Age Research, 8001 Redwood Blvd, Novato, CA 94945, USA

In this introduction I want to summarize where we are in the field and where we are going. What is the future of mitochondrial bioenergetics? A couple of weeks ago I had an idle moment, so I logged on to PubMed and entered the search term 'mitochondria' followed by the years 1950 and 2006, one after the other. The results were fascinating. The numbers of citations per year for mitochondria started off in the bioenergetic prehistory, going back almost 100 years to the first descriptions of mitochondria. For me, the single event that introduced the classic era of mitochondrial bioenergetics was the publication of papers by Chance and Williams in the mid 1950s which first described the oxygen electrode, and described the redox changes of the cytochromes. What happened then was an explosive growth over the next 10 or 15 years in the number of papers, which led to over 3000 papers per year by the time the next revolution came. This was Peter Mitchell's work in the late 1960s. This is interesting: whereas you would expect that the discovery of a mechanism would stimulate a lot of new research, what happened after the three or four years when Peter was publishing these fantastic papers is that the field stagnated for 20 years in terms of numbers of publications. Somehow, because bioenergetics had defined itself so narrowly in terms of understanding how the respiratory chain and ATP synthase works with a little bit of ion transport, this limited the field.

The next explosion of research came when our cell biology colleagues working on cell health and death discovered, sometimes to their discomfort, that mitochondria moved into the centre of the field. In the last 10 years the trend has been almost explosive in terms of the number of papers on mitochondrial bioenergetics. I didn't have time to do a statistical sampling of the different years, but my guess is that 80% of these papers come under the field of mitochondrial physiology: mitochondria in the context of the cell.

What is interesting is where we are going. To quote Donald Rumsfeld, it is the unknown unknowns that will define where the field moves in the next 10 years. In 1994, no one knew that they didn't know how cytochrome c was released, because it wasn't part of the vocabulary. What does this have to do with cell death? It is the things that we don't know that we don't know which are going to define the next 10 years or so.

What Michael Duchen has organized for this meeting is a series of four separate sessions. The first one is mitochondrial 'natural history', a phrase I like because it conjures up images of David Attenborough wandering through the cell and getting excited when he comes close to a mitochondrion. The three papers in this session deal with mitochondrial morphology, fission, fusion, replacement of proteins and shape. Some of the questions that occurred to me and which we will be dealing with are as follows. Do you repair a mitochondrion, or are they scrapped and replaced when they go wrong? How is a damaged mitochondrion recognized, and why doesn't this recognition work well in terms of mitochondrial genetic disease? Do we see more fission than fusion if there are more mitochondria being produced, and what is the relationship between fission/fusion and biogenesis?

The second session is mitochondria and oxidative stress. I'm fascinated by PGC1α, and equally fascinated by novel uncoupling proteins. Does PGC1α regulate gross mitochondrial bioenergetics? Does it regulate the specific induction of antioxidant pathways? What are the novel uncoupling proteins really doing? This is still surprisingly cloudy, some 10 years after the first description of these proteins.

The third session deals with signal transduction. Mitochondrial nitric oxide synthase is still controversial. Mitochondrial Ca^{2+} signalling and endoplasmic reticulum (ER) cross-talk is also a hot topic: what are the conditions where the ER and mitochondria talk to each other, and what are the conditions where they work on their own? Ion channels will be covered: one ion channel that is highly controversial is the mitochondrial K^+-ATP channel. What is the problem with this? Is it present or not? The permeability transition pore is another subject for discussion: when is it important and when is it not? What are the proven cases for its involvement and what are the more speculative cases? With regard to the regulation of cytochrome c release, do we have to look at this separately from the release of AIF and other proteins? Is there a single, holistic mechanism for releasing these proapoptotic proteins from mitochondria or is cytochrome c a special case? Mitochondria and neurodegeneration is an enormous topic: one of the central questions here is why does damage in these different diseases show such tissue specificity? Is it related to the cell or the mitochondrion itself?

The final session looks at mitochondrial mutations. Even with the Pol-g mutations, is the frequency of mitochondrial DNA mutations sufficient to account for the phenotype? That is, is there any disconnect between the proportion of mitochondria that possess the mutation and the dysfunctional phenotype? Why do the mechanisms of autophagocytosis not recognize some damaged mitochondria when there is heteroplasmic coexistence of normal and damaged mitochondria? This seems to work in the β cell; what goes wrong in some mitochondrial diseases where it goes the wrong way, and the mutated mitochondria seem to be dominant over the wild-type ones? Also, why do we start off each generation with perfect mitochondria? What are the mechanisms that protect the germ cell or allow a Darwinian-style selection of perfect mitochondria in each succeeding generation? Finally, where will the next 10 years take us? We clearly don't know the unknown unknowns, but thinking of the known unknowns do we have a feel for where we are going? I feel that part of the future lies in understanding mitochondrial morphology: why are they shaped the way they are, and why do we have exactly the right number in a cell? Why do they appear to go where they are needed?

Outer mitochondrial membrane protein degradation by the proteasome

Albert Neutzner*, Richard J. Youle* and Mariusz Karbowski*†

*Biochemistry Section, SNB, NINDS, NIH Bethesda MD 20892 and †Medical Biotechnology Center, University of Maryland Biotechnology Institute, University of Maryland, 725 W. Lombard St, Baltimore, MD 21201, USA

Abstract. Protein turnover is used for regulatory processes and to eliminate superfluous, denatured or chemically inactivated polypeptides. Mitochondrial proteins may be particularly susceptible to damage induced by reactive oxygen species and several pathways of mitochondrial proteolysis have been illuminated. However, in contrast to matrix and inner mitochondrial membrane protein degradation, little is known about the turnover of integral outer mitochondrial membrane (OMM) proteins or the mechanisms involved. We have found that pheromone treatment of *Saccharomyces cerevisiae* induces the proteasome-dependent elimination of the OMM spanning protein, Fzo1, from the mitochondria and that Fzo1 is ubiquitylated while still associated with the membrane. These characteristic processing steps are similar to those of the endoplasmic reticulum (ER)-associated degradation (ERAD) pathway suggesting the term OMMAD, outer mitochondrial membrane-associated degradation, to describe the process. ERAD is dependent upon ER membrane spanning RING domain E3 ubiquitin ligases suggesting that certain E3 ligases in the OMM may also regulate OMMAD. This led us to clone and characterize all 54 predicted human gene products that contain both RING domains and predicted membrane spanning domains. A surprising number of these localize to mitochondria where some may control OMMAD. Some of these mitochondrial RING domain proteins also regulate mitochondrial morphology, indicating a critical role of ubiquitin signalling in the maintenance of mitochondrial homeostasis.

2007 Mitochondrial biology: new perspectives. Wiley, Chichester (Novartis Foundation Symposium 287) p 4–20

Regulated protein degradation is used to eliminate superfluous, denatured or chemically inactivated polypeptides and in signal transduction. Current evidence shows that in the mitochondria, protein degradation is controlled by pathways compartmentalized in the matrix and the inner mitochondrial membrane. For example, it has been demonstrated that Pim1/Lon proteases, of the AAA protease family homologous to bacterial proteases, degrade excess soluble matrix proteins whereas membrane embedded i- and m-AAA proteases remove integral membrane proteins in the inner mitochondrial membrane. Additional components of this latter system include the chaperone prohibitin and a membrane channel to allow export of the peptide end-products from the matrix to the intermembrane space (Arnold & Langer 2002). Once within the intermembrane space, peptides smaller than 2 kDa would be expected to diffuse freely through the semipermeable outer mitochondrial membrane. In contrast to proteolysis of matrix and inner mitochondrial membrane proteins, little is known about the turnover of outer mitochondrial membrane (OMM) proteins or the molecular components involved.

During our exploration of the regulation of mitochondrial dynamics, we found that the large GTPase of the OMM required for mitochondrial fusion, Fzo1, is degraded during pheromone treatment of yeast cells (Neutzner & Youle 2005). The elimination of Fzo1 from the OMM during pheromone treatment correlates with a decrease in mitochondrial fusion and shorter overall length of mitochondrial tubules. Interestingly, the degradation of Fzo1 was suppressed by the proteasome inhibitor, MG132, indicating that the proteasome system plays a role in OMM protein turnover (Fig. 1). This conclusion has been extended by showing that pheromone-induced Fzo1 degradation depends upon ubiquitin, the proteolytic activity of the 20S proteasomal core particle (*pre1*) and the ATPase subunit (*cim5*) of the 19S regulatory complex of the proteasome (Escobar-Henriques et al 2006). Understanding how the cytosolic proteasomal system could degrade a membrane spanning protein in the OMM comes from other organelles.

In the endoplasmic reticulum (ER), transmembrane proteins are targeted for proteasomal degradation by the endoplasmic reticulum-associated degradation (ERAD) pathway (Meusser et al 2005). This pathway involves: (i) ubiquitylation of the target protein, typically followed by (ii) retrotranslocation of the ubiquitin-tagged protein out of the membrane by AAA ATPases, and (iii) degradation by proteasomes in the cytosol. As Fzo1 spans the OMM twice, proteasomal degradation would likely also require retrotranslocation from the OMM, and therefore participation of a heretofore unknown mitochondrial-associated degradation pathway. Since the OMM may have originated from the phagosomal membrane of a primitive eukaryotic cell that engulfed an endosymbiotic bacteria (yielding the

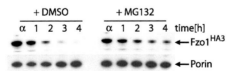

FIG. 1. Pheromone induced Fzo1 degradation is proteasome dependent. Cells of strain yAN018 (*FZO1HA3Δbar1*) were arrested with mating pheromone for 2 h and treated either with 50 μM/ml proteasome inhibitor MG132 (Sigma) in DMSO (1% final concentration) or with DMSO alone as a control. Cells were harvested at 0, 1, 2, 3 and 4 h after MG132 treatment. Protein lysates were analysed by western blot using anti-HA and anti-porin antibodies.

inner mitochondrial membrane [IMM] and matrix of the mitochondrion) (van der Bliek 2000), it is possible that the protein degradation process occurring in the OMM descended from a eukaryotic process that utilizes ubiquitylation. To explore whether degradation of OMM-associated Fzo1 shares other features with ERAD we first determined whether Fzo1 is ubiquitinated.

To attain this, we purified ^{Myc3}Fzo1 by immunoprecipitation from cells with or without additional overexpression of ^{HA3}Ubiquitin. As shown in Fig. 2, immunopurified ^{Myc3}Fzo1 from ^{HA3}Ubiquitin overexpressing cells is modified with ^{HA3}Ubiquitin, while ^{Myc3}Fzo1 from control cells does not show this modification. We therefore conclude that Fzo1 can be tagged with ubiquitin *in vivo*. In order to determine whether Fzo1 is ubiquitylated while on the mitochondrial membrane, we overexpressed ^{Myc3}Fzo1 together with ^{HA3}Ubiquitin and performed subcellular fractionation followed by immunoprecipitation of ^{Myc3}Fzo1 from the heavy membrane fraction. Figure 3A shows that most of the ^{Myc3}Fzo1 sediments were in the heavy membrane fraction (12 K pellet). No Fzo1, however, was detectable in the high speed supernatant (100 K supernatant). Significantly, Fzo1 immunopurified from the heavy membrane fraction was modified with ^{HA3}Ubiquitin, indicating that ubiquitin conjugation of this protein occurs at the OMM. To determine the ubiquitylation site on Fzo1, we analysed several N-terminally truncated Fzo1 mutants for their

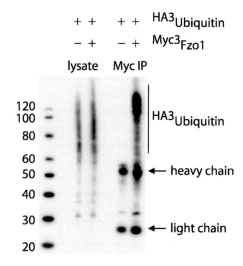

FIG. 2. Fzo1 is ubiquitylated *in vivo*. Cells of strain yAN186 (*GALHA3Ubiquitin*) and yAN251 (*GALHA3Ubiquitin GALMYC3FZO1*) were induced with 2% galactose for 2 h and protein lysates were prepared. Myc3Fzo1 was immunoprecipitated using anti-Myc antibodies (Roche) and then analysed by western blot using anti-HA antibodies to detect HA3Ubiquitin (Roche). The lysate from yAN186 cells served as immunoprecipitation control.

OUTER MITOCHONDRIAL MEMBRANE PROTEIN DEGRADATION

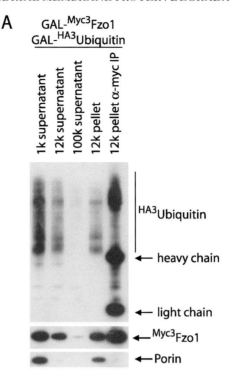

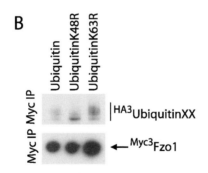

FIG. 3. (A) Fzo1 ubiquitylation takes place on the mitochondrial membrane. Cells of strain yAN251 (*GALHA3Ubiquitin GALMYC3FZO1*) were induced for two hours with 2% galactose to produce Myc3Fzo1 and HA3Ubiquitin. Protein lysates were prepared and fractionated using differential centrifugation at $1000 \times g$ (1K), $12000 \times g$ (12K) and $100000 \times g$ (100K). The heavy membrane fraction (12K pellet) was solubilized and Myc3Fzo1 was immunoprecipitated. The samples were analysed by western blot using anti-HA antibodies to detect HA3Ubiquitin and anti-Myc antibodies to detect Myc3Fzo1. Detection of porin using anti-porin antibodies served as marker for the outer mitochondrial membrane. (B) Cells of strain yAN251 (*GAL-HA3Ubiquitin GALMyc3FZO1*), yAN301 (*GALHA3Ubiquitin GALMyc3FZO1K48R*) and yAN307 (*GALHA3Ubiquitin GALMyc3FZO1K63R*) were induced with 2% galactose and protein lysates were prepared. Ubiquitylation of Myc3Fzo1 was analysed by immunoprecipitation followed by western blot analysis with anti-HA and anti-Myc antibodies.

susceptibility as substrates for ubiquitylation. To this end we overexpressed ^{Myc3}Fzo1 or ^{Myc3}Fzo1 mutants together with ^{HA3}Ubiquitin and performed subcellular fractionation experiments. ^{Myc3}Fzo1 and ^{Myc3}Fzo1 mutant proteins were then immunoprecipitated from the heavy membrane pellet. As shown in Fig. 4, while the deletion

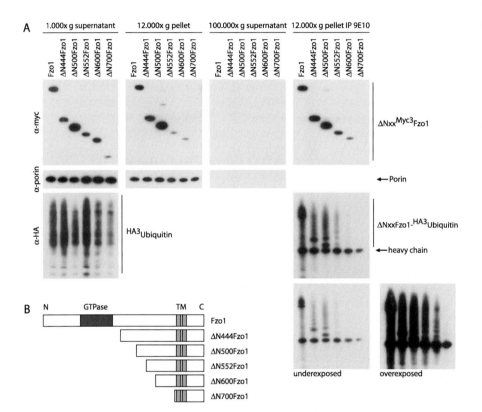

FIG. 4. Analysis of the ubiquitylation of Fzo1 truncation mutants. (A) Cells of strains yAN251 (*GALHA3Ubiquitin GALMyc3FZO1*), yAN231 (*GALHA3Ubiquitin GALMyc3ΔN444FZO1*), yAN227 (*GALHA3Ubiquitin GALMyc3ΔN500FZO1*), yAN216 (*GALHA3Ubiquitin GALMyc3ΔN552FZO1*), yAN294 (*GALHA3Ubiquitin GALMyc3ΔN600FZO1*) and yAN248 (*GALHA3Ubiquitin GALMyc3ΔN700FZO1*) were induced for 2 h with 2% galactose and protein lysates were prepared for subcellular fractionation by differential centrifugation. The heavy membrane fraction (12 K pellet) was solubilized and subjected to immunoprecipitation using anti-Myc antibodies in order to purify Fzo1 truncation mutants. Ubiquitylation of Fzo1 was analysed by western blot using anti-HA antibodies in order to detect HA3Ubiquitin. The detection of porin served as marker for the mitochondrial outer membrane and as a control for the subcellular fractionation. In order to show the different ubiquitylation levels of the Fzo1 mutants three different exposures ('underexposed' and 'overexposed') are shown. (B) Fzo1 constructs used in Figure 4A. GTPase marks the GTPase region of Fzo1. TM marks the two transmembrane regions which anchor Fzo1 in the outer mitochondrial membrane. ΔNxxx stands for an N-terminal deletion of xxx amino acids.

of amino acids 1 to 500 did not impair the localization of Fzo1 to the heavy membrane fraction, any further deletion greatly diminished the mitochondrial localization of Fzo1. Deletion of the first 552 and 600 amino acids still allowed mitochondrial import to a small extent whereas the deletion of amino acids 1–700 significantly inhibited protein expression and localization to heavy membranes. The analysis of heavy membrane localized Fzo1 truncation mutants by immunoprecipitation showed that even the deletion of the first 600 amino acids of Fzo1 still allowed ubiquitylation to occur. These results show a number of potential ubiquitylation sites exist on Fzo1 consistent with a role of ubiquitylation machinery that specifically targets substrates in the outer mitochondrial membrane.

Targeting proteins for proteosomal degradation involves the formation of a ubiquitin chain formed by lysine 48-linked ubiquitin moieties, rather than lysine 63-linked ubiquitin moieties that are formed during regulatory processes. To address whether Fzo1 is modified with lysine 48 or lysine 63 ubiquitin chains, we overexpressed ^{Myc3}Fzo1 together with ^{HA3}UbiquitinK48R or ^{HA3}UbiquitinK63R (Fig. 3B). Analysis of ^{Myc3}Fzo1 ubiquitylation by immunoprecipitation showed that overexpression of ^{HA3}UbiquitinK48R greatly inhibited the formation of ubiquitin chains while overexpression of ^{HA3}UbiquitinK63R did not diminish ubiquitin chain formation compared to ^{HA3}Ubiquitin. This suggests that Fzo1 is modified with lysine 48-linked ubiquitin chains, consistent with a role of the proteasome in degradation of Fzo1. Together, these data show that turnover of Fzo1 might be initiated by ubiquitylation at the OMM followed by proteasomal degradation in the cytosol. The molecular mechanism that mediates retrotranslocation of Fzo1 from the OMM to the proteasome as well as the nature of the E3 ligase for Fzo1 ubiquitylation on mitochondrial membranes remain elusive. As discussed above degradation of Fzo1 in *S. cerevisiae* at the OMM can include (i) ubiquitylation of Fzo1 and (ii) proteasome-dependent elimination of this membrane spanning protein. These processing steps are similar to those of ERAD, suggesting that an analogous pathway, which we term outer mitochondrial membrane associated degradation (OMMAD), may participate in protein quality control of the OMM.

One key step in elimination of ERAD substrates is the ubiquitylation of the target proteins by ER transmembrane E3 ligases, including Hrd1 and Doa10 in yeast (Carvalho et al 2006) and gp78 in mammals (Song et al 2005). To investigate potential E3 ligase components involved in OMMAD we first attempted to determine whether any RING-finger E3 ubiquitin ligases localize to mitochondria. We turned to mammalian systems where ERAD components have been extensively studied. The human genome encodes about 345 RING domain proteins based on NCBI gene and Genbank searches using several C3HC4 RING motifs of known E3-ubiquitin ligases related to $CX_2CX_{[9,39]}CX_{[1,3]}HX_{[2,3]}CX_{[2]}C_{[4,48]}CX_{[2]}C$ as reference sequences. Of these 345 candidates, 54 different cDNAs have predicted membrane-spanning domains as revealed by the transmembrane helix prediction program TMHMM (*http://www.cbs.dtu.dk/services/TMHMM*). We cloned the cDNAs of these 54 genes from a human brain cDNA library or commercially available clones assuming that some of them may localize to mitochondria and participate in OMMAD. The domain organization of proteins cloned in our screen is shown in Fig. 5. A remarkable array of transmembrane domain organizations

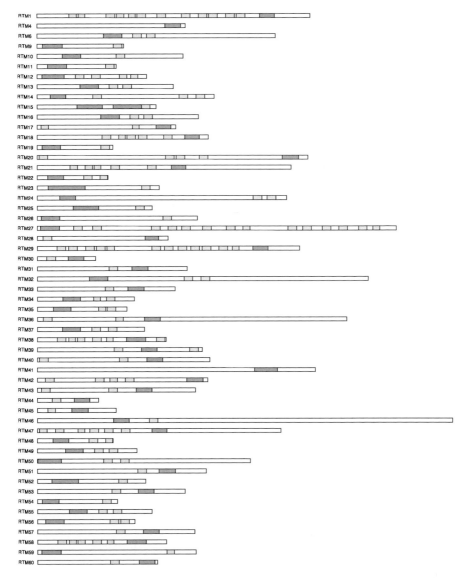

FIG. 5. Domain organization of 55 human RING domain containing proteins. A dark grey box marks the RING domains and lighter grey boxes mark putative transmembrane regions.

appears with predicted membrane spanning regions ranging from a single C-terminal anchor to 14 membrane spanning domains suggesting a wide diversity of membrane events that may control protein stability through the ubiquitin pathway.

To analyse the subcellular localizations of our candidate membrane-associated E3 ubiquitin ligases, cDNAs of all 54 genes were subcloned into mammalian expression vectors in frame with yellow fluorescent protein (pYFP), which we used as a fluorescent reporter in protein localization studies. To avoid potential subcellular localization artefacts due to fusion with the 27 kDa YFP, we constructed both N-and C-terminal chimeras (resulting in: 'YFP-protein of interest' and 'protein of interest-YFP' orientations). Figure 6 presents an overview of the subcellular localizations of these RING domain proteins. Importantly, we found that of the 54 integral membrane RING domain proteins, nine localized to mitochondria. Six of these mitochondrial proteins had the RING domain N-terminal to the membrane spanning regions and three displayed C-terminal RING domains, presumably orienting the RING domains toward the ubiquitination machinery in the cytosol. None of the nine mitochondrial-localized RING domain proteins displayed a mitochondrial import consensus sequence based on the PSORTII prediction program (*http://psort.nibb.ac.jp/form2.html*) consistent with OMM localizations. The number of membrane spanning segments ranged from 1 to 5. One of these mitochondrial associated proteins contains a RING IBR RING motif found in a subfamily of RING proteins that includes Parkin, an E3 ligase required for

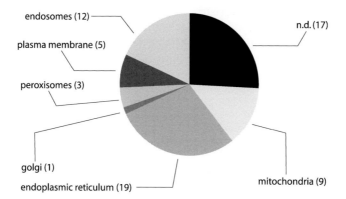

FIG. 6. Cellular distribution of membrane anchored RING domain proteins. Fifty-four open reading frames coding for proteins containing a RING domain as well as predicted transmembrane domains were fused to YFP and expressed in HeLa cells. The localization of these YFP fusion proteins was analysed by fluorescence microscopy and confirmed by co-staining with standard organelle markers. An overview of the various membrane organelles which harbour transmembrane RING domain proteins revealed by our screen is shown. Localization 'not determined (n.d.)' refers to work in progress.

mitochondrial maintenance that is mutated in certain familial cases of Parkinson's disease. To explore the potential role of these RING domain proteins in mitochondrial protein turnover, we mutated conserved histidines in the RING domain, already shown to be critical for this domain function, in all nine proteins to inactivate their potential E3 ligase activity. Current studies will determine if these mitochondrial RING domain proteins participate in the degradation of the human homologues of Fzo1, Mfn1 and Mfn2, or other integral OMM proteins. As discussed below, one of the nine proteins found by us to localize to mitochondria has been recently shown to participate in ubiquitin- and proteasome-mediated self-degradation, two characteristics predicted for OMMAD.

Interestingly, of the nine mitochondrial RING proteins identified in our screen, only MARCH5 was associated with this organelle in a previous proteomic study (Taylor et al 2003). The absence of prior identification of eight mitochondrial RING domain proteins (Mootha et al 2003, Taylor et al 2003) may reflect the poor recovery of the OMM in highly purified mitochondrial preparations, relative to the abundance of the inner mitochondrial membrane, or the difficulty in identifying proteins with several transmembrane domains by mass spectrometry procedures. It is also possible that a low abundance or tissue specific expression of the mitochondrial RING domain proteins could complicate their detection in proteomic studies. Recently, two groups have identified a role of MARCH5 in ubiquitylation of OMM substrates, including Fis1 and Drp1, two proteins vital for mitochondrial fission (Nakamura et al 2006, Yonashiro et al 2006). Interestingly, one of these proteins, the large GTPase Drp1, co-immunoprecipitates with MARCH5 only in an ubiquitylated form. These data also support a role of MARCH5 in the mitochondrial ubiquitin signalling. In addition, MARCH5 seems to be itself ubiquitylated and degraded by the proteasome, sharing two characteristics with another OMMAD substrate, Fzo1.

Discussion

Mitochondria produce reactive oxygen species that can be very damaging for mitochondrial proteins, lipids and DNA. Thus, selective removal and degradation of denatured and oxidized proteins is likely to be an important process in mitochondrial quality control. In contrast to the matrix and inner mitochondrial membrane, where systems of proteolysis are well characterized, protein turnover mechanisms in the OMM are not known. We propose, based upon examples of integral OMM protein ubiquitylation and proteasome degradation, that a process similar to ERAD regulates protein turnover in the OMM and call this outer mitochondrial membrane associated degradation or OMMAD.

To explore the mechanism of OMM degradation we cloned and localized all the membrane spanning RING domains identified in the human genome. Confocal localization of YFP fusion proteins identified nine membrane spanning RING domain proteins on mitochondria. We will determine whether these are involved in OMM protein degradation. In addition to these candidate E3 ligases, E3 ligases not integrally spanning membranes may also regulate OMM protein stability. *MDM30* has been found in yeast to be required for mitochondrial fusion (Fritz et al 2003). Interestingly, *MDM30* contains an F-box domain frequently found in proteins complexed with the cytosolic SCF family of E3 ligases. Although not required for pheromone induced Fzo1 degradation, *MDM30* is involved in steady state degradation of Fzo1 and thereby controls the process of mitochondrial fusion but surprisingly through a proteasome independent process (Escobar-Henriques et al 2006). Mfb1, another F box protein, also regulates mitochondrial morphology (Durr et al 2006). Other examples of cytosolic E3 ligase regulation of mitochondria include Rsp5 and Parkin. The soluble HECT domain E3 ligase, Rsp5, is required for mitochondrial distribution and morphology in yeast although the substrate for this ubiquitin ligase activity involved in mitochondrial regulation remains unknown (Fisk & Yaffe 1999). The RING IBR RING domain protein, Parkin, not predicted to have membrane spanning domains is localized in the cytosol and associated with a variety of organelles (Kubo et al 2001) including mitochondria (Kuroda et al 2006) and is involved in mitochondrial responses to oxidative damage in metazoans. Importantly, Parkin is mutated in certain cases of early onset, familial Parkinson's disease (Kitada et al 1998) and may disrupt the normal participation of OMMAD of damaged mitochondrial proteins. The most likely mitochondrial substrates of soluble E3 ligases would reside either embedded in the OMM or bound to proteins on the OMM. However, analogous to ERAD extraction of soluble ER luminal proteins, inter-membrane space proteins in mitochondria are also candidate substrates for proteosomal degradation, particularly via the transmembrane OMM E3 ligases defined in this work.

References

Arnold I, Langer T 2002 Membrane protein degradation by AAA proteases in mitochondria. Biochim Biophys Acta 1592:89–96
Carvalho P, Goder V, Rapoport TA 2006 Distinct ubiquitin-ligase complexes define convergent pathways for the degradation of ER proteins. Cell 126:361–373
Durr M, Escobar-Henriques M, Merz S, Geimer S, Langer T, Westermann B 2006 Nonredundant roles of mitochondria-associated F-box proteins Mfb1 and Mdm30 in maintenance of mitochondrial morphology in yeast. Mol Biol Cell 17:3745–3755
Escobar-Henriques M, Westermann B, Langer T 2006 Regulation of mitochondrial fusion by the F-box protein Mdm30 involves proteasome-independent turnover of Fzo1. J Cell Biol 173:645–650
Fisk HA, Yaffe MP 1999 A role for ubiquitination in mitochondrial inheritance in Saccharomyces cerevisiae. J Cell Biol 145:1199–1208
Fritz S, Weinbach N, Westermann B 2003 Mdm30 is an F-box protein required for maintenance of fusion-competent mitochondria in yeast. Mol Biol Cell 14:2303–2313

Kitada T, Asakawa S, Hattori N et al 1998 Mutations in the parkin gene cause autosomal recessive juvenile parkinsonism. Nature 392:605–608
Kubo SI, Kitami T, Noda S et al 2001 Parkin is associated with cellular vesicles. J Neurochem 78:42–54
Kuroda Y, Mitsui T, Kunishige M et al 2006 Parkin enhances mitochondrial biogenesis in proliferating cells. Hum Mol Genet 15:883–895
Meusser B, Hirsch C, Jarosch E, Sommer T 2005 ERAD: the long road to destruction. Nat Cell Biol 7:766–772
Mootha VK, Bunkenborg J, Olsen JV et al 2003 Integrated analysis of protein composition, tissue diversity, and gene regulation in mouse mitochondria. Cell 115:629–640
Nakamura N, Kimura, Y, Tokuda M, Honda S, Hirose S 2006 MARCH-V is a novel mitofusin 2- and Drp1-binding protein able to change mitochondrial morphology. EMBO Rep 7:1019–1022
Neutzner A, Youle RJ 2005 Instability of the mitofusin Fzo1 regulates mitochondrial morphology during the mating response of the yeast Saccharomyces cerevisiae. J Biol Chem 280: 18598–18603
Song BL, Sever N, DeBose-Boyd RA 2005 Gp78, a membrane-anchored ubiquitin ligase, associates with Insig-1 and couples sterol-regulated ubiquitination to degradation of HMG CoA reductase. Mol Cell 19:829–840
Taylor SW, Fahy E, Zhang B et al 2003 Characterization of the human heart mitochondrial proteome. Nat Biotechnol 21:281–286
van der Bliek AM 2000 A mitochondrial division apparatus takes shape. J Cell Biol 151:F1–4
Yonashiro R, Ishido S, Kyo S et al 2006 A novel mitochondrial ubiquitin ligase plays a critical role in mitochondrial dynamics. EMBO J 25:3618–3626

DISCUSSION

Nicholls: There seems to be a relationship between the excision of these proteins and the morphological changes. What is the link that triggers the morphological change as the result of protein excision?

Youle: From what we know now, I'd say there are two different links. One is that in yeast the fusion protein Fzo1 (Mfn in mammals) is excised and degraded by the proteasome, and this promotes fission because the mitochondria can't fuse. We have seen this physiologically in yeast during mating. It makes sense that they might want to have less interconnected mitochondria when the cells mate, because cells of different mating types fuse, and then undergo meiosis and may profit from a nice mixture of their mitochondria in the daughter cells. What we are now seeing is that other E3 ligases in mammals (and this is more speculative) function in a regulatory process of mono-ubiquitination that appears not to function by inducing proteasomal degradation. The ubiquitin can be added and taken off by de-ubiquitinylating enzymes in a regulatory capacity. We are looking to see whether Drp1 dynamics on the membrane are modulated by a monoubiquitination process. We know very little of what regulates Drp1 accumulation at mitochondrial scission sites and how it winds around mitochondria, and how it cycles off these helices.

Heidi McBride has already reported that Drp1 is sumoylated, but you wouldn't think from the sequences that the E3 ligases we have identified are sumo ligases. I should also acknowledge that Heidi McBride has identified but not yet published one of the nine E3 ligases on mitochondria that we found.

Jacobs: Have you excluded the possibility that as well as co-localization with Drp1, there is another target? You presented a co-localization in the Drp1 dots, but there could be, from what you have said, another target that is being ubiquitinated and destroyed. Although it is a different family of proteins, this reminds me of what is going on in the nuclear cell cycle at the metaphase–anaphase transition. Cyclins are suddenly degraded, and this makes the mitotic process irreversible. Similarly, when a mitochondrial network is about to undergo fission, the last thing you want to happen is that the process will get half way and then get stuck, such that the sealing events won't work properly and there will be leakage and loss of membrane integrity. Potentially, this could be a system for ensuring that the event, once a trigger has been given and a signal received, would then be irreversible. I would predict from this that there would be other targets.

Youle: I hope that you are right. A couple of other proteins have been reported to co-localize with Drp1 in the spots. One of these, my favourite protein Bax, co-localizes in these sites during apoptosis. This is where we are going with studies of the ring domain proteins on cytochrome C release processes during apoptosis. Also, some of the mitofusins may also assemble at these Drp1 foci: we will hear later that perhaps fission and fusion can occur rapidly, perhaps at the same spot. There may be a bit of back and forth before they decide to drift apart. This can be seen in confocal experiments in other people's mitochondrial scission studies. We have localized Mfn2, another large GTPase in the dynamin family, in foci with Drp1. E3 ligases are also involved in degradation of Mfns in yeast and likely in mammals.

Shirihai: One clear difference between a nucleus going through mitosis and the mitochondria going through fission is that in the latter, multiple fission events occur in an unsynchronized fashion. There is therefore an interesting additional complexity to do with the localization of the fission protein to the mitochondria that go through fission and not to other mitochondria.

Rizzuto: One of the key aspects is correlating rapid physiological events with fusion and fission. We want to understand how the mitochondrial morphology decodes rapid events, and copes with the rapidity and complexity of signalling. Do you think this can be a system for rapidly translating inputs arriving to the cell surface into mitochondrial morphological changes?

Youle: Absolutely. When a molecule spans the membrane nine times, this is likely to be not just a localization motif but also as a sensory motif in the membrane for detecting something. We can look to the ER for examples, such as sterol binding in the membrane, which regulates E3 ligases. There are probably a lot of lipid interactions with these proteins. How does a membrane know that it needs more phosphatidyl serine? Perhaps there are sensors in the membrane through which E3 ligases can regulate transcriptional events to make more phosphatidyl serine synthase.

Lemasters: In your work with GFP-Bax, Bax translocated to mitochondria and formed aggregates that were next to mitochondria rather than truly in the organelle. These diffuse bodies always struck me as some kind of aggresome. Might proteins that are being extracted from mitochondria be going into that aggresomal structure? Does Bax get ubiquinated, and are there other chaperone proteins/HSP proteins in association with this potential aggresome?

Youle: Perhaps you are right. We have done a lot of work looking at Bax translocation to mitochondria, and it forms these focal spots that co-localize with Drp1 and mitochondrial scission spots. Looking under the microscope in real time, we can see spots of Bax on tips of mitochondria and the mitochondria get shorter and shorter. Perhaps a unifying hypothesis might be that it is involved in this type of mitochondrial disintegration process. You can see where I'm going: the ER can retrotranslocate lumenal proteins. There is a lot of retrotranslocation of cytochrome c during apoptosis through a pore or another unknown mechanism, perhaps by a process related to ERAD.

Spiegelman: Can you put in perspective the fission/fusion process in terms of the need for mitochondrial respiration in various tissues? If your muscle has a need for an increased acute or chronic level of mitochondrial respiration, where does this fission/fusion process fit into the ordinary requirement for more oxidative metabolism?

Youle: There is one paper by Capaldi and colleagues (Rossignol et al 2004) in which they showed that when cultured cells were switched from a glucose to a galactose containing medium, the mitochondria become longer in order to drive oxidative metabolism. We have confirmed this. There are probably direct links between mitochondrial fusion, network elaboration and mitochondrial biogenesis.

Nicholls: Is that an increase in the mass of mitochondria or an inhibition of fission?

Youle: It is probably both.

Nicholls: How does the biogenesis fit into the cycle?

Youle: It's like the currency markets. We all spend money and earn money (that's like fission/fusion) but as the population and economy grows the government prints more money (this is biogenesis).

Spiegelman: I'd be shocked if there's no intelligence in terms of adding currency. I assume that the mitochondria must know how to add currency. In the liver which is in a fasting response, or a muscle in response to exercise, the bells and whistles might be quite different.

Halestrap: You suggested that the mitochondrial E3 ligases might ubiquitinylate proteins that aren't membrane bound, like transcription factors. Is there any evidence that this can occur, that it doesn't have to be a membrane protein but can be a soluble protein?

Youle: It could be both ways. There could be membrane-spanning ligases that regulate soluble targets that regulate events in the nucleus. With the unfolded protein response, there is a membrane-spanning protein that can directly regulate splicing of a transcription factor in the cytosol that increases expression and thereby nuclear gene transcription. It is quite likely that there is an entire class of soluble E3 ligases that can impact on and regulate the mitochondrial proteins.

Halestrap: This would potentially create a link between biogenesis and fission/fusion.

Reynolds: How do you think about the rate of turnover of these ubiquitinated proteins in relation to mitochondrial proteins in general, on the outer or inner membrane or even in the matrix? Do you have benchmarks for things that turn over slowly or stay put? When you see something disappear in 6 h is that fast compared with everything else?

Youle: We haven't looked at this systematically. From the examples I have shown, we can compare the E3 ligase mutant that cannot self-ubiquitinate with the wild-type form: one is very stable after 36 h and the latter is gone in three hours. With our yeast work, Fzo1 in the absence of α factor is quite stable.

Reynolds: Can you make the distinction between turning over a single protein versus turning over an entire organelle?

Youle: This goes back to how much do we 'retrofit' an old mitochondrion? This is teleological, but if you were a cell and you had an inactive OMP that has been oxidized by ROS, would one denatured protein hurt your mitochondrion? How many denatured OMPs could a mitochondrion tolerate? Then, at what point do you pitch in the whole mitochondrion and make a new one?

Giulivi: I have a comment on the turnover. I was shocked by the turnover of your proteins: it was so fast. It has been shown that the whole mitochondrion has a half-life of 10 d, and then for individual proteins, anything from 1–7 days. Your thesis is that it's actually just a few hours.

Youle: One proviso about what I presented is that these are all overexpressed proteins. Perhaps they are not wanted anyway and are purposely degraded, whereas the wild-type proteins are more stable.

Martinou: You showed that mitofusin 1 is degraded by the proteosome. When you overexpress it and inhibit the proteosome, there is more Mfn1. So I was expecting to see ubiquitinated Mfn1 appearing, and this didn't seem to be the case.

Youle: Usually it is a small subset that is ubiquitinated. When proteosome inhibitors are added usually it is a fairly small pool of Mfn1 that is ubiquitinated. We typically IP the protein of interest and then immunostain for ubiquitin on that protein. This picks up a few percent of the bound ubiquitin. Regulatory mono-ubiquitination results in a more discrete new band of the protein at a higher molecular weight.

Jacobs: One of the things mentioned in the early part of the discussion is that the particular example you started with in yeast is actually a signalling pathway involved in the function of the gamete. This cell is about to undergo fusion with another cell. In this case the rationale for fragmentation of the network is clear: it would facilitate the segregation of individual mtDNA variants that have arisen. When the mitochondria fuse after the cell fusion has been initiated, each such variant in the population could undergo a genetic testing by recombination with all the variants in the other cell. This makes clear genetic sense. However, while we were discussing this I recalled the observation from Schatten's group (Sutovsky et al 1999, 2000) that in mammalian sperm there is ubiquitination of one or more sperm mitochondrial proteins, including prohibitin. I don't know quite what to make of this. It hints at a mechanism of turnover that is involved in doing the opposite, preventing the fusion of mitochondria at fertilization, because for whatever reason paternal inheritance is to be excluded. One obvious way of ensuring this would be to render the sperm mitochondria incapable of fusion by eliminating some protein on their surface.

Schon: I don't buy this because we don't get the elimination in interspecific hybrids, and there is no reason why the ubiquitination should not operate in these.

Jacobs: Unless there is specificity of the protein that is directing the ubiquitin ligase. Sutovsky's results actually suggest this, since ubiquitination seems to be defective in interspecies hybrids.

Lemasters: There are some recent studies that show that ubiquitination targets sperm mitochondria for possible autophagic degradation in fertilized eggs. This is a process that actively eliminates male gamete mitochondria and their mtDNA.

Jacobs: That is presumably quite a slow process. It doesn't happen immediately after fertilization. If there is an errant mitochondrion present in the zygote, and it is going to be turned over at some point, that is fine. But if you can prevent it talking to any of the other mitochondria present, in the same way that in the events of fertilization there are fast and slow blocks to polyspermy, this could represent a 'fast' and a 'slow' backup mechanism to prevent paternal inheritance and recombination.

Shirihai: This suggests a possible mechanism for the observation that mitochondria that are leaving a fission event are relatively depolarized and prevented from going through another fusion event (unpublished data). This raises the question of whether this kind of ubiquitination is happening rapidly on mitochondria that are relatively damaged, preventing them from going through another fusion unless they can recover by themselves, thus not contaminating the rest of the population with damaged material.

Youle: One thing we think about along these lines is why are fission and fusion so dynamic. If there are damaged proteins on mitochondria, another way to get rid of them would be to selectively group them on a tip, divide the tip off and then send that to the autophagosome, enriched in bad material.

Scorrano: Do you think that ubiquitination of mitochondria could be the signal for targeting them to the autophagosomes?

Youle: I hope not.

Nicholls: The problem with that is that you can only detect what is happening on the outer membrane, whereas the bioenergetic deficit is more likely to be on the inner membrane. So how does the autophagosome know that something has gone wrong two membranes away?

Jacobs: In your screening, you just assayed for targeting to mitochondria. You haven't verified in every case that the targeting is unique to the outer membrane. My question is therefore are any of these ring domain proteins targeted to the inner membrane, which then would be involved in an even more outrageous process where they were extracting proteins from the inner membrane that were damaged and then were exporting them somewhere to be turned over?

Youle: Then you would need to get ubiquitin into the inner membrane space. It will be quite a bit of work to address this because we would need antibodies to all nine proteins.

Scorrano: We have some data showing that Opa1 can be ubiquitinated. Perhaps there is something happening inside, or it could be that there is a small soluble fraction.

Spiegelman: I can imagine it is difficult to do the experiment where a cell has so-called 'good' and 'bad' mitochondria, and taking it to an extreme physiological state such as denervating muscle, where half the muscle mass is going to be turned over in a week, including the mitochondria which will be eaten by autophagy. If you want to chew up mitochondria, this is an extreme way to do it, and it is real physiology. Muscle can expand or contract depending on energetic need, so if you cut the sciatic nerve you can watch the mitochondria getting chewed up.

Nicholls: What is fascinating about this is that it is not a mitochondrial 'dysfunction', but the mitochondria not being fully utilized. They are sitting there in state IV because there is not the ATP demand. Is that a signal?

Spiegelman: Biochemically, the system sounds a lot more accessible than some of these other models. You'll set off a process of mass mitochondrial autophagy by denervating the nerve.

Nicholls: How is a mitochondrion that isn't being fully utilized recognized? Does it have a slightly higher membrane potential?

Youle: There is a well established pathway for peroxisome autophagy, which can be selectively induced by metabolic changes.

Scorrano: In co-operation with Marco Sandri we are inducing fission in muscle and looking at what happens to the muscle fibres. Fission is by itself one of the most prominent atrophic signals. Somehow it triggers atrophy of the fibres. This is even more powerful than induction of FoxO or other atrophic factors.

Spiegelman: How do you know that by stimulating the fission you are not introducing some dysfunction?

Scorrano: We measured it and found no mitochondrial dysfunction.

Halestrap: I wonder whether the so-called contact sites between the inner and outer membranes might be relevant, whereby the matrix, in response to energy status or damage, can talk to outer membrane proteins and thus affect their ubiquitination. Just a thought.

References

Rossignol R, Gilkerson R, Aggeler R, Yamagata K, Remington SJ, Capaldi RA 2004 Energy substrate modulates mitochondrial structure and oxidative capacity in cancer cells. Cancer Res 64:985–93

Sutovsky P, Moreno RD, Ramalho-Santos J, Dominko T, Simerly C, Schatten G 1999 Ubiquitin tag for sperm mitochondria. Nature 402:371–372

Sutovsky P, Moreno RD, Ramalho-Santos J, Dominko T, Simerly C, Schatten G 2000 Ubiquitinated sperm mitochondria, selective proteolysis, and the regulation of mitochondrial inheritance in mammalian embryos. Biol Reprod 63:582–590

PA-GFP: a window into the subcellular adventures of the individual mitochondrion

Sarah E. Haigh, Gilad Twig, Anthony A. J. Molina, Jakob D. Wikstrom, Motti Deutsch and Orian S. Shirihai

Department of Pharmacology and Experimental Therapeutics, Tufts University School of Medicine, 136 Harrison Ave, Boston, MA 02111, USA

Abstract. Mitochondrial connectivity is characterized by matrix lumen continuity and by dynamic rewiring through fusion and fission events. While these mechanisms homogenize the mitochondrial population, a number of studies looking at mitochondrial membrane potential have demonstrated that mitochondria exist as a heterogeneous population within individual cells. To address the relationship between mitochondrial dynamics and heterogeneity, we tagged and tracked individual mitochondria over time while monitoring their mitochondrial membrane potential ($\Delta\Psi_m$). By utilizing photoactivatible-GFP (PA-GFP), targeted to the mitochondrial matrix, we determined the boundaries of the individual mitochondrion. A single mitochondrion is defined by the continuity of its matrix lumen. The boundaries set by luminal continuity matched those set by electrical coupling, indicating that the individual mitochondrion is equipotential throughout the entire organelle. Similar results were obtained with PA-GFP targeted to the inner membrane indicating that matrix continuity parallels inner membrane continuity. Sequential photoconversion of matrix PA-GFP in multiple locations within the mitochondrial web reveals that each ramified mitochondrial structure is composed of juxtaposed but discontinuous units. Moreover, as many as half of the events in which mitochondria come into contact, do not result in fusion. While all fission events generated two electrically uncoupled discontinuous matrices, the two daughter mitochondria frequently remained juxtaposed, keeping the tubular appearance unchanged. These morphologically invisible fission events illustrate the difference between mitochondrial fission and fragmentation; the latter representing the movement and separation of disconnected units. Simultaneous monitoring of $\Delta\Psi_m$ of up to four individual mitochondria within the same cell revealed that subcellular heterogeneity in $\Delta\Psi_m$ does not represent multiple unstable mitochondria that appear 'heterogeneous' at any given point, but rather multiple stable, but heterogeneous units.

2007 Mitochondrial biology: new perspectives. Wiley, Chichester (Novartis Foundation Symposium 287) p 21–46

Mitochondria in living cells form a dynamic reticulum that undergoes continuous remodelling through fusion and fission events; this process is termed mitochondrial dynamics (Chan 2006, Karbowski & Youle 2003). Mitochondrial dynamics have been shown to influence diverse mitochondrial functions. Increased mitochondrial fusion, induced by overexpression of fusion proteins, Mitofusin 1 and 2 (Mfn1, 2) or OPA1 (Chen et al 2003), or knockdown of fission proteins Drp1 or Fis1 (Lee et al 2004), prevents mitochondrial induced apoptosis. Reduced mitochondrial fusion (caused by knockdown of Mfn1/2 or OPA1) has been shown to affect mitochondrial metabolism by decreasing OXPHOS capacity and reducing mitochondrial membrane potential ($\Delta\Psi_m$) (Bach et al 2003, Chen et al 2003, 2005). These effects are thought to result from an inability of mitochondria to complement each other through networking.

How connected is the mitochondrial web?

Many cell types do not have the elaborate, mitochondrial networks that have been described in fibroblasts, HeLa cells and COS-7 cells (De Giorgi et al 2000). In fact, several observations suggest that the ramified tubules of the above cells may not represent a network with interconnected lumens, but rather a collection of juxtaposed mitochondria. In concert with this theory, functional studies of $\Delta\Psi_m$ and mitochondrial calcium uptake have shown high levels of mitochondrial heterogeneity in nearly every cell type observed (Collins et al 2002). These conflicting observations are reconciled if $\Delta\Psi_m$ is asymmetrically distributed across an individual mitochondrion. Indeed, studies of the longitudinal voltage profile of a single mitochondrial unit have found significant gradients. Using the $\Delta\Psi_m$-sensitive dyes tetramethylrhodamine methyl ester (TMRM) (Diaz et al 2000) and JC1 (5,5′,6,6′-tetrachloro-1,1′,3,3′-tetraethylbenzimidazolyl carbocyanine iodide) (Bereiter-Hahn 1990), the voltage gradient along a given mitochondrion was measured to be at least 15 mV, and usually greater than 30 mV. In contrast, evidence of low electrical resistance and free molecular diffusion across the mitochondrial matrix predicts that the organelle be equipotential (Partikian et al 1998).

These findings emphasize the importance of experimentally defining the term 'individual mitochondrion'. Electron microscopy (EM) has been used to generate three-dimensional reconstruction images of mitochondria. Although it provides high resolution data, EM is not compatible with functional measurements and cannot follow changes over time. Confocal microscopy, along with fluorescent probes, has made simultaneous monitoring of mitochondrial morphology, $\Delta\Psi_m$ and mitochondrial calcium uptake possible. However, defining the boundaries of a single mitochondrion using mitochondrial fluorescent probes alone is nearly impossible in cells with all but the sparsest populations of mitochondria.

The study presented here takes advantage of our ability to trace the lumen of an individual mitochondrion using a photoactivatible form of GFP, and focuses on the units that constitute a mitochondrial web. We suggest that the approach utilized in this study may provide insights into the physiological significance of mitochondrial fusion and fission.

Defining an 'individual mitochondrion', an experimental approach

Utilizing photoactivatible GFP (PA-GFP)

PA-GFP was developed in the lab of Jennifer Lippincott-Schwartz and has been used to follow individual organelles including mitochondria (Patterson & Lippincott-Schwartz 2002).

We fused the targeting peptide of ABCme, a mitochondrial inner membrane transporter, to the N-terminus of PA-GFP, thereby targeting it to the mitochondrial matrix: this construct is known as PA-GFP$_{mt}$. Cells were transduced with a lentivirus encoding PA-GFP$_{mt}$ and allowed to express and accumulate the protein for 48 hours. The mitochondria were subsequently labelled with the membrane potential sensitive probe TMRE prior to imaging.

Transition of PA-GFP$_{mt}$ to its active (fluorescent) form was achieved by photo-isomerization using a Zeiss Camelion 2-photon laser (750 nm) to give a 375 nm photon-equivalence at the focal plane. During photo-activation of a distinct branch of the mitochondrial web, the PA-GFP$_{mt}$ molecules in that region irreversibly isomerized to their active, fluorescent state. This enabled selective activation of regions smaller than 0.5 μm^2. In the absence of photo-activation, PA-GFP$_{mt}$ molecules remained stable in their inactive form. Using the multi-track scanning mode of Zeiss LSM-510 microscope, red-emitting TMRE was excited with a 543 nm helium/neon laser, and its emission recorded through a 650–710 nm filter. Activated PA-GFP$_{mt}$ protein was excited using a 488 nm argon laser at intensities that are 10 times lower than those used to detect the inactive protein. Emission was recorded through a BP 500–550 nm filter.

The advantage of using a 2-photon laser for large non-flat cells

While the photolabelling of mitochondria in fibroblasts, COS-7 and HeLa cells can be achieved with reasonable resolution without a 2-photon laser, photolabelling of mitochondria in non-flat cells, with spherical or elevated structures poses an additional challenge. Mitochondria of spherical cells (such as blood cells) may be positioned at the same XY location but in different focal planes. A conventional laser would activate the PA-GFP$_{mt}$ of all mitochondria that are in the laser path. The use of a 2-photon laser enables us to target individual units in the Z axis and prevent damaging mitochondria that are not photolabelled.

Tracking mitochondria in non-adherent cells

We utilized a novel transparent honeycomb shaped scaffold to follow fusion events in non-adherent cells (Deutsch et al 2006) (Fig. 1). The cell array corrals non-adherent cells within micron-sized wells, and prevents cell movement. Using this

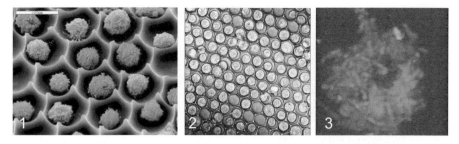

FIG. 1. Photolabelling individual mitochondria in non adherent cells using 2-photon laser. Each well of the cell array measures 20 μm in diameter, and can accommodate one cell. (1) SEM micrograph of Jurkat T-lymphocytes. Scale bar, 20 μm. (2) Brightfield image of K562 erythroleukemic cells within the cell array. (3) A single K562 cell labelled with both PA-GFP$_{mt}$ and TMRE. The activated PA-GFP$_{mt}$ reveals a single mitochondrion. Reprinted from Deutsch et al 2006 with permission. See original publication for colour version of figure.

approach we were able to observe non-adherent cells over extended periods of time, obtain high resolution images, and photoactivate individual mitochondria using a 2-photon laser, without the cell moving and inadvertently activating multiple mitochondria.

Lessons from tracking individual mitochondria

Fusion, a mixing of matter

- *An individual mitochondrion can be defined by lumenal continuity:* The diffusion boundaries within each mitochondrial network were revealed when the isomerized PA-GFP$_{mt}$ equilibrated across the lumen, a process that took 1–5 seconds. The network boundaries were defined by the continuity of its matrix space, within the larger mitochondrial web (Figs. 2A, B). Interestingly, the continuous shape of the web as revealed by TMRE alone, does not reliably predict the boundaries

FIG. 2. Fusion is revealed by a continuous lumen. (A) COS-7 cell expressing PA-GFP$_{mt}$ prior to photoactivation. The cell is co-stained with TMRE highlighting all mitochondria. Scale bar, 20 μm. (B) A magnified section of the cell shown in (A). Images of the mitochondrial web before, and after one and two photoactivation steps (images 1–3, respectively). The regions targeted for photoactivation are indicated (white squares). Scale bar, 2 μm. (C) Lateral diffusion of photoactivated PA-GFP$_{ABCme}$ identifies segments with continuous inner membrane. The square indicates the location where photoactivation was performed. Dashed lines indicate the mitochondrial branch that was highlighted by inner membrane continuity. (D) Fusion occurred between 210 and 225 seconds, resulting in dilution of PA-GFP$_{mt}$ and reduction in its fluorescence intensity (FI). Note that the spread of PA-GFP$_{mt}$ to the fusing mitochondrion (indicated by the arrow) was instantaneous and did not leave TMRE-stained areas in which PA-GFP$_{mt}$ was below noise. (E) A sequence of fusion and fission events of an individual mitochondrion in an INS-1 cell. Fusions occurred, resulting in dilution of PA-GFP$_{mt}$ and reduction in its FI. Reprinted from Twig et al 2006 with permission. See original publication for colour version of figure.

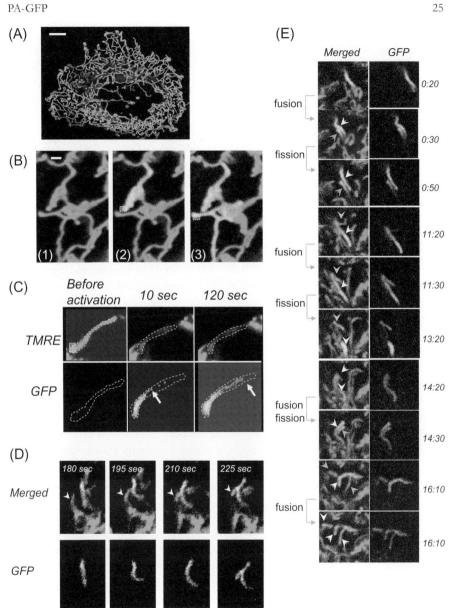

of a network (Fig. 2B). This phenomenon was observed in all cell types examined, including INS-1 ($n = 37$), COS-7 ($n = 23$) and K562 cells ($n = 10$).

- *Inner membrane proteins diffuse during fusion:* In addition to the matrix targeted PA-GFP$_{mt}$, we created a construct in which PA-GFP was fused to the entire ABCme protein, thereby restricting it to the inner mitochondrial membrane. In comparison to PA-GFP$_{mt}$, diffusion of PA-GFP$_{ABCme}$ from the photolabelled mitochondria into its fusion mate was slow (1–2 minutes versus 5 seconds for the matrix protein) (Fig. 2C). This is mainly due to the slower diffusion of the inner membrane protein within the membrane of the individual mitochondrion. The time it took for diffusion to reach a steady state was 60–300 seconds for PA-GFP$_{ABCme}$ ($n = 30$), compared to 1–5 seconds for PA-GFP$_{mt}$ ($n = 300$). Although slower, PA-GFP$_{ABCme}$ diffused throughout the entire mitochondrial segment that showed electrical continuity as described for the matrix PA-GFP.

- *Fusion is a transient event:* Most fusion events proceeded in a manner similar to the one presented in Fig. 2D. Two mitochondria became juxtaposed. After a period of time with no transmission of PA-GFP$_{mt}$ (1 second to several minutes), the PA-GFP$_{mt}$ passed between the two mitochondria revealing lumenal continuity. Within 5 minutes of fusion the two mitochondria were likely to undergo a subsequent fission event, dividing at the site where fusion previously occurred (Fig. 2E). Similar patterns have been observed in plant mitochondria, where it was referred to as 'kiss and run'. Our data suggest that since membrane diffusion also occurs during these events, this term may not accurately describe fusion events.

- *Fusion can occur at the tip or the middle of mitochondria:* Tracking individual units enabled us to determine the fusion site. At least one third of the events occurred at a site which is remote from the poles of a mitochondrion. Examples are shown in Fig. 2.

Tracking $\Delta\Psi_m$ in individual mitochondria

Our approach provides a means to identify the boundaries of a single mitochondrion within the complex mitochondrial architecture. Furthermore, it enables us to track changes in $\Delta\Psi_m$ through consecutive fusion and fission events.

Unlike TMRE, PA-GFP$_{mt}$ fluorescence is insensitive to changes in mitochondrial membrane potential. Therefore, the ratio quotient of the fluorescent intensities (FI) of TMRE and PA-GFP$_{mt}$ was employed to quantify small changes in $\Delta\Psi_m$ while taking into account potential shifts in focal plane (Fig. 3). This technique has similar accuracy to scanning multiple sections in the z-plane to optimize the focal plane. However, it has two advantages due to the fact that it requires only one imaging plane: (1) improved time resolution and (2) minimized laser exposure thus extending the time available for the experiment and minimizing photo damage.

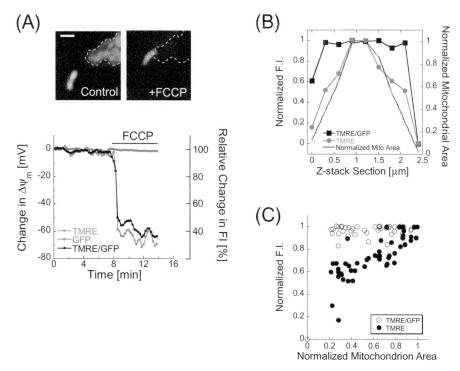

FIG. 3. Defining mitochondrial lumen continuity and $\Delta\Psi_m$ using TMRE and PA-GFP$_{mt}$. (A) PA-GFP$_{mt}$ fluorescence is insensitive to changes in mitochondrial membrane potential. Addition of 1 μM of FCCP induced $\Delta\Psi_m$ depolarization indicated by the attenuated TMRE FI (dashed line) but did not influence the FI of PA-GFP$_{mt}$. FI of TMRE and PA-GFP$_{mt}$ are given in percentage change from their value at time zero. The quantitative changes in the ratio of TMRE/GFP were converted to $\Delta\Psi_m$ and are shown over time. Scale bar, 2μm. (B) Use of TMRE/GFP ratio corrects for focal plane changes. The dependency of TMRE FI and the TMRE/GFP ratio on the image focal plane is shown for a single mitochondrion that was imaged at nine sections in the z-plane (intervals of 0.3μm). The mitochondrial area captured at each focal plane was derived from the green signal and is normalized to the slice of maximal area. (C) A summary of 12 experiments was used to identify the conditions at which TMRE/GFP ratio is insensitive to focal plane changes (dot plot). Each mitochondrion was imaged at 3–9 sections in the z-plane capturing different portions of its maximal area. Reprinted from Twig et al 2006 with permission. See original publication for colour version of figure.

- *Examining subcellular heterogeneity over time.* It has been shown repeatedly that mitochondria are functionally heterogeneous. By tagging two individual mitochondria and following them simultaneously, we were able to address the following question: 'Does the observed heterogeneity represent dynamic changes in the membrane potential of individual mitochondria or static heterogeneity over time?'

Utilizing the TMRE/GFP ratio approach, we followed two to three individual mitochondria at a time in primary β cells. PA-GFP$_{mt}$ was used in this experiment to track the $\Delta\Psi_m$ of photolabelled mitochondria and to determine if a fusion event had occurred during the imaging period. From these data we constructed a membrane potential time chart for each mitochondrion. A representative chart is shown in Fig. 4 demonstrating the stability of the membrane potential of the individual mitochondrion as well as stability of the difference between two mitochondria.

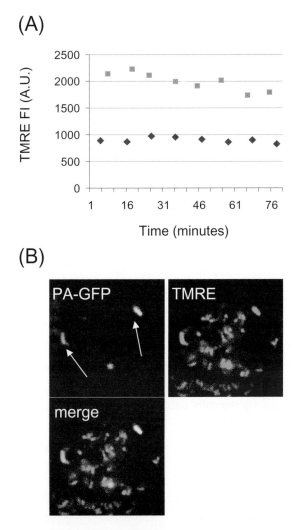

FIG. 4. Tracking the membrane potential of two individual mitochondria in a single cell over time. The $\Delta\Psi_m$ of two individual mitochondria in an individual primary β cell expressing adenoviral encoded PA-GFP$_{mt}$ was monitored using TMRE and PA-GFP$_{mt}$. (A) Traces of $\Delta\Psi_m$ of the two mitochondria. (B) Image showing the monitored mitochondria. The mitochondria maintained a $\Delta\Psi_m$ difference of ~10 mV throughout the imaging.

Membrane potential charts representing single mitochondria from 17 different INS-1 cells demonstrate that, at least during the interphase between fission and fusion events, the membrane potential remains within ±–1.7 mV of the baseline for 80% of the time. These data support the hypothesis that subcellular heterogeneity represents a collection of mitochondria with different but stable membrane potentials.

- *The mitochondrial unit: lumen continuity and uniform membrane potential.* We compared the diffusion boundaries of activated PA-GFP$_{mt}$ and PA-GFP$_{ABCme}$ to the boundaries determined by electrical coupling. Two types of experiments were conducted to validate the reliability of PA-GFP in defining the network lumen boundaries. Both discount the possibility that matrix structures permit electrical coupling but block protein diffusion (Fig. 5).

Individual branches of the mitochondrial web in COS-7 cells expressing PA-GFP$_{mt}$ were targeted with a 2-photon laser at a photo-toxic intensity. Ten seconds after photoactivation the structure of the network matrix was revealed by the spread of PA-GFP$_{mt}$ molecules. One minute after the application of the high-dose laser the mitochondrial membrane potential collapsed and TMRE fluorescence disappeared from the electrically coupled segment (Fig. 5A). The electrically coupled segment and the area defined by PA-GFP$_{mt}$ diffusion were colocalized in COS-7 ($n = 15$), INS-1 cells ($n = 10$) and K562 cells ($n = 5$). Similar results were obtained with PA-GFP$_{ABCme}$. We conclude that an individual mitochondrion, as defined by lumenal continuity, has homogeneous membrane potential, i.e. is electrically coupled throughout its length.

Monitoring $\Delta\Psi_m$ identifies fission events in which the two mitochondria remain juxtaposed

Immediately after photoactivation, PA-GFP$_{mt}$ identifies a continuous matrix lumen that is equipotential and permits analysis of its size and morphology. A spontaneous fission event that separates this lumen into two uncoupled parts can be identified by the onset of a voltage gradient and can be morphologically demonstrated by a second photoactivation (Fig. 5B).

The possibility that the mitochondrial inner membrane might experience fission events while the outer membrane maintains its integrity was suggested in yeast mitochondria by Jakobs et al (2003). They noted an intermittent distribution of matrix volume inside tubules of outer membranes. We demonstrate here that even in the absence of intermittent matrix morphology, it is possible for the matrix to be compartmentalized in distinct, but juxtaposed units. This is consistent with the hypothesis that mitochondria undergo independent fission of the inner membrane and suggests that fission of the outer and of the inner membranes might involve

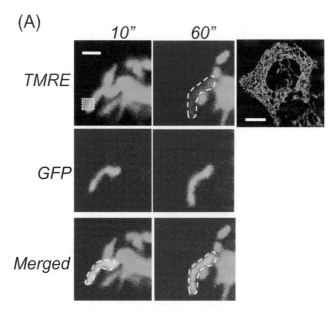

FIG. 5. Intra-network electrical synchronization and dissociation occur without morphological interruption. (A) Mitochondrial membrane depolarization is revealed by a decrease in TMRE FI, yet does not affect the FI of activated PA-GFP$_{mt}$. The ratio between the FI of TMRE and PA-GFP can be used to track changes in mitochondrial membrane potential over time. A COS cell was stained with TMRE (red, shown here as darker grey) and transfected with PA-GFP (green, shown here as lighter grey). A mitochondrion, highlighted by the dashed line, depolarizes at 60 seconds, resulting in decreased TMRE FI. The merged image depicts changes in the ratio between TMRE FI and PA-GFP$_{mt}$ FI. (B) Intra-network electrical dissociation, in the absence of morphological changes, is accompanied by a diffusion block in the matrix lumen. Two-step photo-activation experiment is shown. Step 1: identification of a network. In a schematic illustration of the mitochondrial network (upper row), the site of the first photoactivation (dashed square) and ROI #1 and #2 from which membrane potential was measured (circles) are illustrated. Images were acquired 10 seconds after photoactivation. Pseudo-colour image depicts the FI of GFP$_{mt}$. Step 2: A second photoactivation was applied when a voltage difference of more than 5 mV was measured between ROI (Region of Interest) #1 and #2. Location of the second photoactivation is marked in the schematic illustration (top). PA-GFP$_{mt}$ activated in the second stage redistributed asymmetrically between ROI #1 and #2, indicating discontinuity of their matrices. The boundaries of the disconnected matrices were plotted in the merged and green-signal-only images (dashed white line). Calibration bar, 2 µm. Graph: the change in $\Delta\Psi_m$ over time measured from the two ROIs. Events '1' and '2' indicate the timing of first and second photoactivations. Note that ~4.5 min following the first photoactivation (arrow) the two traces diverge and that during this voltage difference, the lumen continuity between ROI #1 and #2 was altered as shown above. Reprinted from Twig et al 2006 with permission. See original publication for colour version of figure.

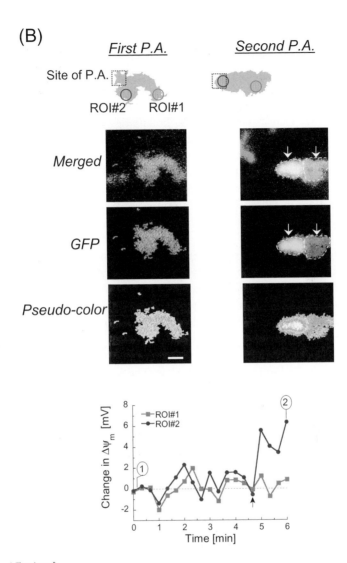

FIG. 5. (*Continued*)

different molecular machineries (Malka et al 2005). We find that networked mitochondrial webs are actually composed of multiple units arranged within submicron proximity. Interestingly, in all cell types studied, each of the lumenally continuous units constituted a similar 1–2% fraction of the entire web with occasional units being doubled or tripled in size. This result demonstrates that cells with different mitochondrial connectivity may have similar unit size but different spatial arrangements.

Fission and fragmentation

While the test for lumenal continuity determines the boundaries of a certain network at the time of photoactivation, electrical coupling can be continuously monitored and can report on possible dissociations that may occur over time. To follow the connectivity dynamics of the individual network, we used the PA-GFP$_{mt}$ approach to first identify and label the boundaries of each network, and then follow its connectivity by monitoring electrical coupling and synchronicity within the network. While electrical coupling was found in the first 30 seconds in all networks tested (the maximal difference within a network was 1.13 mV ± 1.45 mV, $n = 20$), some networks demonstrated electrical dissociation at later time points. Such dissociations in $\Delta\Psi_m$ occurred without any apparent gross change in the inner membrane morphology or location of the mitochondrion, as judged from the TMRE image.

This observation illustrates the differences between mitochondrial fission and mitochondrial fragmentation. The latter is a result of mitochondrial fission of both inner and outer mitochondrial membranes followed by mitochondrial movement—three processes that are energy-dependent (Legros et al 2002). Fission, in contrast, reflects an alteration in the continuity of the lumen regardless of mitochondrial movement.

This implies that proteins that govern fission events might be overlooked if fragmentation is used to identify such agents. The need to distinguish fragmentation from fission is exemplified by the proteins Fis1 and Drp1. While both proteins mediate fission, they are associated with different patterns of mitochondrial movement and spatial arrangement, cytosolic spread versus perinuclear clustering (Frieden et al 2004, Szabadkai et al 2004).

The voltage difference between daughter mitochondria provides an insight into the metabolic changes that accompany a fission event. Although the mechanism that underlies this voltage difference is unclear, these results suggest that $\Delta\Psi_m$-dependent metabolic processes, such as ATP synthesis and calcium uptake, can be distributed unequally between daughter mitochondria, at least within the time scale of several minutes after a fission event. Further studies are needed to elucidate the biophysical asymmetry following fission events.

Addressing morphological and physiological discrepancies

We have shown here that a mitochondrion can be defined by the continuity of its lumen, the length of which is equipotential. Other works that have measured a voltage gradient along a single mitochondrion have arrived at different conclusions. Our results suggest two possible reasons for this discrepancy: (1) the problems associated with identification of the boundaries of a mitochondrial unit, especially with two juxtaposed mitochondria; (2) the occurrence of morphologically imperceptible fission and fusion events.

Submicron proximity and the overall appearance of the mitochondrial web are parameters that reflect the gross fusion/fission balance and were suggested to be of functional significance for the propagation of calcium waves (Szabadkai et al 2004) and reactive oxygen radicals (ROS) (Aon et al 2004). However, defining a mitochondrial unit by visual (microscopical) inspection alone may not prove sufficiently accurate in predicting the boundaries of lumen continuity. In the experiments reported here, the mitochondrial web as illuminated by TMRE fluorescence was too ramified and interconnected to allow visual decomposition into mitochondrial units (Fig. 1). A string of aligned but individual mitochondria is superficially identical to an elongated mitochondrion with a continuous matrix. In addition, networks may interconnect in unpredictable configurations. Under confocal microscopy examination, even in an adherent and flat cell such as COS-7 or INS-1, a given branch in a mitochondrial unit may be luminally continuous with a relatively distant part of the web. When the boundaries of a mitochondrion are not accurately known, $\Delta\Psi_m$ measured from loci of two adjacent mitochondria can be falsely interpreted as differences along a single mitochondrion.

The unpredictable changes in lumen continuity further complicate identification of mitochondrial units. We show here that morphologically imperceptible fissions may occur within the mitochondrial unit and alter lumen continuity (Fig. 5). Since these events may result in significant voltage differences between the two adjacent daughter mitochondria, a perceived voltage gradient within an apparently single mitochondrion may be misrepresentative of luminal separation.

The observation of silent fission and the accompanying changes in $\Delta\Psi_m$ (Fig. 5) also sheds light on a fundamental controversy in mitochondrial physiology. There is a growing body of evidence showing that a mitochondrial web acts in a coordinated manner as judged from the spread of electrical signals (De Giorgi et al 2000), ROS (Aon et al 2004), calcium waves (Szabadkai et al 2004) and apoptosis-inducing factors (Pacher & Hajnoczky 2001). In contrast, a separate set of works has demonstrated that mitochondria exist as discrete units and encompass a wide range of $\Delta\Psi_m$ (Collins et al 2002, Collins & Bootman 2003, Twig et al 2004). In the context of our findings, the large differences in $\Delta\Psi_m$ between adjacent mitochondria do not contradict the function of the mitochondria as co-operative networks. Our approach demonstrates that while the mitochondrial web may be composed of discrete units, the dynamic dimension of this web is turning it into an interconnected network on an intermittent basis.

Reconciling heterogeneity with dynamic networks

Mitochondria within the individual cell have been found to be heterogeneous in every parameter tested so far. This includes calcium concentration, membrane potential and morphology. However, photoconversion of PA-GFP$_{mt}$ in the lumen of 20–30% of mitochondria in a cell results in equilibration across the entire mitochondrial web within 30–50 minutes (Karbowski et al 2004), suggesting a very high rate of exchange of material. Heterogeneity is surprising given the extensive level of networking reported.

A number of explanations should be considered. (1) Fusion events do not result in diffusion and equilibration of matrix and membrane components thereby leaving the two fused units unchanged. This has been ruled out by the experiments described above in which an inner membrane molecule (PA-GFP$_{ABCme}$) is tracked through fusion events and shows diffusion and distribution across the mitochondrial web through fusion events. The time it takes for equilibration of an inner membrane protein to occur is longer than that for matrix soluble molecules. (2) Each mitochondrion is frequently changing its properties with time. If changes in membrane potential among the population of mitochondria within the cell are not synchronized, it could result in the observation of overall heterogeneity at any single snapshot. This possibility has been ruled out, as the differences in $\Delta\Psi_m$ remained permanent over time (Fig. 4). (3) Not all mitochondria participate in the dynamic process of networking resulting in the generation of a subpopulation of mitochondria with different bioenergetic and electrophysiological parameters. This would imply that defective mitochondria might not be capable of complementation or rescue by active mitochondria. To test this possibility we have imposed damage through laser radiation and monitored mitochondria that became depolarized. In a typical response to laser damage, mitochondria go through $\Delta\Psi_m$ flickering and depolarization that is then followed by fragmentation. This fragmentation is not reversed by fusion suggesting that, at least during the short period of the imaging (1–2 hours), the damaged unit is not being rescued through fusion and complementation.

These results suggest the possibility that damaged mitochondria might be segregated from the mitochondrial network, possibly in preparation for removal by mechanisms such as autophagy. A combination of fission that is followed by selective fusion can form a segregation mechanism by which dysfunctional units are sorted prior to being removed. Together, these can form a filter mechanism that functions as a quality control for the mitochondrial population.

Reference list

Aon MA, Cortassa S, O'Rourke B 2004 Percolation and criticality in a mitochondrial network. Proc Natl Acad Sci USA 101:4447–4452

Bach D, Pich S, Soriano FX et al 2003 Mitofusin-2 determines mitochondrial network architecture and mitochondrial metabolism. A novel regulatory mechanism altered in obesity. J Biol Chem 278:17190–17197
Bereiter-Hahn J 1990 Behavior of mitochondria in the living cell. Int Rev Cytol 122:1–63
Chan DC 2006 Mitochondrial fusion and fission in mammals. Annu Rev Cell Dev Biol 22:79–99
Chen H, Detmer SA, Ewald AJ, Griffin EE, Fraser SE, Chan DC 2003 Mitofusins Mfn1 and Mfn2 coordinately regulate mitochondrial fusion and are essential for embryonic development. J Cell Biol 160:189–200
Chen H, Chomyn A, Chan DC 2005 Disruption of fusion results in mitochondrial heterogeneity and dysfunction. J Biol Chem 280:26185–26192
Collins TJ, Bootman MD 2003 Mitochondria are morphologically heterogeneous within cells. J Exp Biol 206:1993–2000
Collins TJ, Berridge MJ, Lipp P, Bootman MD 2002 Mitochondria are morphologically and functionally heterogeneous within cells. EMBO J 21:1616–1627
De Giorgi F, Lartigue L, Ichas F 2000 Electrical coupling and plasticity of the mitochondrial network. Cell Calcium 28:365–370
Deutsch M, Deutsch A, Shirihai O et al 2006 A novel miniature cell retainer for correlative high-content analysis of individual untethered non-adherent cells. Lab Chip 6:995–1000
Diaz G, Falchi AM, Gremo F, Isola R, Diana A 2000 Homogeneous longitudinal profiles and synchronous fluctuations of mitochondrial transmembrane potential. FEBS Lett 475:218–224
Frieden M, James D, Castelbou C, Danckaert A, Martinou JC, Demaurex N 2004. Ca(2+) homeostasis during mitochondrial fragmentation and perinuclear clustering induced by hFis1. J Biol Chem 279:22704–22714
Jakobs S, Martini N, Schauss AC, Egner A, Westermann B, Hell SW 2003 Spatial and temporal dynamics of budding yeast mitochondria lacking the division component Fis1p. J Cell Sci 116:2005–2014
Karbowski M, Youle RJ 2003 Dynamics of mitochondrial morphology in healthy cells and during apoptosis.Cell Death Differ 10:870–880
Karbowski M, Arnoult D, Chen H, Chan DC, Smith CL, Youle RJ 2004 Quantitation of mitochondrial dynamics by photolabeling of individual organelles shows that mitochondrial fusion is blocked during the Bax activation phase of apoptosis. J Cell Biol 164:493–499
Lee YJ, Jeong SY, Karbowski M, Smith CL, Youle RJ 2004 Roles of the mammalian mitochondrial fission and fusion mediators Fis1, Drp1, and Opa1 in apoptosis. Mol Biol Cell 15: 5001–5011
Legros F, Lombes A, Frachon P, Rojo M 2002 Mitochondrial fusion in human cells is efficient, requires the inner membrane potential, and is mediated by mitofusins. Mol Biol Cell 13:4343–4354
Malka F, Guillery O, Cifuentes-Diaz C et al 2005 Separate fusion of outer and inner mitochondrial membranes. EMBO Rep 6:853–859
Pacher P, Hajnoczky G 2001 Propagation of the apoptotic signal by mitochondrial waves. EMBO J 20:4107–4121
Partikian A, Olveczky B, Swaminathan R, Li Y, Verkman AS 1998 Rapid diffusion of green fluorescent protein in the mitochondrial matrix. J Cell Biol 140:821–829
Patterson GH, Lippincott-Schwartz J 2002 A photoactivatable GFP for selective photolabeling of proteins and cells. Science 297:1873–1877
Szabadkai G, Simoni AM, Chami M, Wieckowski MR, Youle RJ, Rizzuto R 2004 Drp-1-dependent division of the mitochondrial network blocks intraorganellar Ca2+ waves and protects against Ca2+-mediated apoptosis. Mol Cell 16:59–68

Twig G, Graf SA, Wikstrom JD et al 2006 Tagging and Tracking Individual Networks within a Complex Mitochondrial Web Using Photoactivatable GFP. Am J Physiol Cell Physiol 291: C176–C184

DISCUSSION

Schon: Have you tried photoactivatable GFP targeted to the outer membrane?
Shirihai: No.
Schon: The reason I ask is that if you do the much maligned JC1, in a normal cell you will see green spaghetti with orange pearls on the string. As an electrician, I find it hard to understand this because it means that the membrane potential is not uniformly distributed. The only explanation that makes sense is a sausage kind of model where the outer membranes are fused but the inner membranes are not, or are delimited, as you showed.
Shirihai: There are some data from the work of Manuel Rojo that mitochondrial fission of the inner membrane can go separately from that of the outer membrane. This is one possibility. There are other data from Nunari, which argued that the outer-only fusion is an unstable intermediate.
Concerning the beads on string that is seen in images made with the mitochondrial dye JC1, JC1 images should be interpreted with much caution as this dye is capable of generating a large array of artefacts. Nevertheless, these artefacts are very artistic and colourful. PA-GFP diffusion through the matrix perfectly overlaps with the segments that are electrically coupled, based on TMRE staining. These segments do not have the beads on string arrangements, a finding that suggests that JC1 appearance as beads on string is a type of artefact.
Lemasters: Lan Bo Chen, who described a lot of these dyes, showed that JC1 forms red aggregates, called J aggregates (Smiley et al 1991). These are precipitates. If you do confocal microscopy at high resolution you can see these aggregates within the homogenous matrix of simple spherical mitochondria, such as those of hepatocytes (Lemasters & Ramshesh 2007). These precipitates are responsible for the red fluorescence. Many of the other potential-indicating cationic fluorophore dyes do the same, except that the aggregates do not fluoresce, which distinguishes JC1 from others. In a sinuous mitochondrion, when you see these spots of JC1 red fluorescence, it does not imply differences in electrical potential within a single mitochondrion. The red spots just show the physical aggregates of the dye, which increase as more dye is accumulated with a more negative membrane potential.
Nicholls: There are a number of papers suggesting that mitochondria in neurites possess a higher membrane potential than mitochondria in cell bodies (e.g. Overly et al 1996) However, a problem with JC1 is that it is the slowest permeating of the different 'TMRM' class of indicators since it has chlorides in the structure. This means that the rate at which it loads depends on the geometry of the cell. It loads

more rapidly into thin neurites than it does into large globular cell bodies, resulting in artefacts.

Rizzuto: We all agree that JC1 doesn't work well, but Eric Schon's point is correct: simply by looking with fluorescence microscopy we see connectivity, and then we see diffusion. This either means that there are separate entities and they are still there without actually fusing, or that the OM is fused and the IM is not. The experimental tool to address this is an OM targeted photoactivatable GFP. We have an OM GFP; it is just a matter of making it photoactivatable. I suspect his answer is correct.

Hajnoczky: Rojo and colleagues have a paper (Malka et al 2005) that describes separation of the outer membrane fusion from the inner membrane fusion based on drug sensitivity. We have expanded on this work using photoactivated fluorescent proteins targeted to various submitochondrial compartments. Using both intermembrane space and matrix targeted probes, we find that the breakdown of the inner membrane barrier is tightly coupled to the breakdown of the outer membrane barrier. The gap between the two events is 2 seconds or so. Thus, the outer and inner membrane fusion events are closely co-ordinated with each other in healthy cells. Interestingly, the mixing of an outer membrane-associated probe often precedes the mixing of the intermembrane space targeted probe by several seconds. This would indicate that some interactions are first confined to the mixing of outer membrane associated factors. This may or may not be followed by a complete fusion event.

Youle: It has been known for a long time that chloroplasts can form stable outer mitochondrial networks called stromules that connect separated matrix compartments. We see many situations where it is possible to have that intermediate state. Thinking about changes in fluorescence intensity with changes in membrane potential, imagine if you had thin mitochondria with a knob at the end. Depending on your z stacking of sections or the plane of focus, you could see greater fluorescence simply due to a change in mitochondrial thickness.

Halestrap: That would be true for any dye.

Nicholls: This would be controlled for in Orian Shirihai's experiments because he is looking ratiometrically between the GFP and the TMRM.

Halestrap: This assumes that GFP is equally expressed in every part. It probably is, but this is an assumption.

Bernardi: Probes all have their own problems, particularly the MitoTrackers. As Luca Scorrano and I showed many years ago (Scorrano et al 1999), these are powerful inhibitors of complex I. I would be concerned with long-term experiments using MitoTracker. JC1 is bad, but MitoTrackers are not much better in long-term experiments.

Martinou: In our hands, if the inner membrane is not able to fuse, the outer membrane never fuses.

Scorrano: We never try to measure outer membrane fusion when we inhibit inner membrane fusion. We always try to measure lumenal continuity. We all know that the IMM is impermeant, but the IMM fuses with another IMM during mitochondrial fusion. In general, during fusion of membranes there are pores. So what happens when an IMM fuses with another IMM? What happens to mitochondrial membrane potential during this fusion process? Is there a pore or is there another mechanism, which would be a revolution in membrane fusion because any membrane that fuses inside the cell opens a pore?

Nicholls: In other words we would have a dramatic depolarization event. Is that so?

Bernardi: If there is continuity you would gain the membrane potential immediately. I don't see how you could measure it in that timescale.

Nicholls: It is only if there is leakage to the cytoplasm that we would have problems.

Can we talk about the heterogeneous membrane potential after fusion? This was a central theme of Orian Shirihai's paper. Orian, you find your membrane potentials to be clamped and stable before a fusion event, whereas Ian Reynolds and others have reported dramatic fluctuations in membrane potential in different systems.

Reynolds: It is one of those things that has been examined in a number of different cell types. We started doing this in primary neurons in culture. We have seen the same kind of thing in primary astrocytes in culture, and immortalized HT22 hippocampal cells. The question arose of whether we were measuring real physiology, as opposed to an artefact with illumination of rhodamine dyes spinning off some oxidants and triggering pathophysiological events. I'm not sure we robustly put this aside.

Nicholls: The greatest sensitivity for phototoxicity is when you are doing the sorts of experiments that Orian is doing with a ×100 objective and a small laser spot.

Shirihai: I didn't show it here, but we performed extensive studies to determine the safe window for the use of the specific laser beams for these experiments. We specifically verified that the laser intensity used would not compromise membrane potential during the experiments for up to 2 h and that the cell is not responding with flickering of mitochondrial membrane potentials leading to depolarization. With TMRE and laser illumination it is easy to end up with flickering potentials. Such artefacts may lead to the conclusion that membrane potential is very unstable.

Brand: I'm addressing the issue of whether the mitochondrial membrane potential is stable or flickers. We need to remember that all cells are not the same. In particular, you are using β cells, which have bizarre energetics. Of all the cells these are the ones where membrane potential is controlled by substrate supply. In other cells, such as neurons or muscle cells, substrate supply will not control the mitochondrial membrane potential so strongly. You may have a unique situation in β cells that isn't applicable to other cells where internal regulatory pathways are running.

Shirihai: We also did this work in COS cells and found the same thing.

Bernardi: I am worried that there might be a misconception here. In this field it is assumed that depolarization is a bad thing, but it shouldn't be forgotten that the more active mitochondria (which are synthesizing ATP) have a lower membrane potential. If your mechanism were true it would discard the best mitochondria. How do you know that this depolarization is a bad signal? Have you tried to add oligomycin to see whether there is repolarization? It can't be just the membrane potential: you need to have something else that sorts out whether the depolarization is a sign of initial disease or a signal for autophagy, or just the physiological response to increased ATP demand.

Nicholls: In this country we have something called the National Grid, with a few hundred power stations feeding into the same grid that is in synchrony. In the same way the cell has a cytoplasmic ATP/ADP pool, with mitochondria contributing to this grid. As soon as a mitochondrion depolarizes then the ATP synthase will reverse and pull up that membrane potential by utilizing the ATP/ADP generated by the cell. Along similar lines, oligomycin causes a greatly increased heterogeneity of membrane potentials when it is added. Have you done this?

Shirihai: When we talk about depolarization, it is better if we use numbers. If we measure the depolarization/hyperpolarization in β cells responding to glucose, we never see values above 7–8 mV. Similarly, reduction in glucose from 8 mM to 4 mM does not lead to depolarization that exceeds 10 mV. On the other hand when we examine cells under oxidative stress, for example under glucotoxicity, we see increased heterogeneity and we observe the appearance of damaged mitochondria with a membrane potential that is 20 mV from the average of the cell. Previous studies in isolated mitochondria suggest that a reduction of 14 mV should lead to a 10-fold decrease in ATP synthesis capacity. Concerning the possibility that the depolarized daughter mitochondrion has a reduced membrane potential due to increased ATP synthase activity, we tested whether oligomycin can prevent the appearance of depolarized daughter mitochondria during the fission event, and it does not. That is, when a depolarized mitochondrion is generated by a fission event, this is not the result of increased ATP synthase activity, but more likely due to decreased respiration or a leak. We also tested if oligomycin can prevent the recovery of those daughters that go through recovery and the result was negative, indicating that the recovery phase is not due to ATP hydrolysis.

O'Rourke: We are getting into issues of how the mitochondrial network communicates. There are two possibilities: they could be physically connected electrically, or there could be coupling via diffusible factors. One of the things I've been studying for years is oscillations in the mitochondrial network, and how these can be a physiological mechanism for synchronization as well. We published a paper in the *Biophysical Journal* that looked at cardiac cells to see whether there are small oscillations in membrane potential and ROS production that couple mitochondria and synchronize the network. In heart cells, the main form of synchrony seems to be through local interactions of diffusible messengers. A lot depends on the spatial organization of the mitochondria: the packing and how close they are. In the cardiac cell they are essentially a cubic lattice. In cells that you are studying, you may get different behaviour in the closely packed mitochondria around the nucleus versus the long spindly mitochondria. Part of this can be due to ROS-dependent coupling. When these mitochondria are depolarizing they are probably producing a burst of ROS as well. If neighbouring mitochondria are close by, they may be affected by this and it could set off a chain of events. Flickering is also a physiological control mechanism in the mitochondrial network: we see correlated behaviour in the network that can be modified by changing the respiration rate or ROS production rate.

Martinou: Are the UCPs 2 and 3 involved in this process?

Brand: It has occurred to me as well: it is an interesting possibility, but there is no information about this.

Adam-Vizi: How evenly distributed are the fusion proteins in the daughter mitochondria, having polarized and depolarized membrane potential? Is it possible that in the population of mitochondria that are depolarized, the number of fusion proteins is lower than in the original mitochondria and in those with higher membrane potential? There have been some publications (Bach et al 2003, Chen et al 2005) indicating that the amount of fusion proteins and the membrane potential of mitochondria are related.

Shirihai: We didn't look at the amount of fusion proteins in two daughter mitochondria. In relation to Richard Youle's paper, it's possible that a depolarized daughter mitochondrion is one that is being targeted by ubiquitination.

Jacobs: Can you clarify one point. You showed that mitochondria that have a lower potential after fission take longer to undergo the next fusion event.

Shirihai: Yes, it is longer, or never occurs at all.

Jacobs: So how dark does a mitochondrion have to get, such that it is never allowed to fuse again? Do you see dark mitochondria, with the stain that you are using, which can still fuse? It is a simple probabilistic issue?

Shirihai: Lower membrane potential mitochondria would have lower probability of fusing. In mitochondria with very low or no detectable TMRE we didn't see fusion events happening. This doesn't mean that these mitochondria will never be able to recover.

Halestrap: We haven't really answered Richard's criticism of measuring membrane potential. Surely, if there are mitochondria of different depth, even within the resolution of confocals, a deeper mitochondrion will give a bigger fluorescent signal even though the membrane potential is the same. I don't see how we can get accurate estimates of 5 mV difference of membrane potential. When we hear that oligomycin doesn't raise the potential, this makes me even more worried. How can we be sure we are getting an accurate measurement?

Nicholls: We have an answer to this. If you look at a single mitochondrion, this will go in and out of the focal plane. The question is whether we see a heterogeneous apparent membrane potential within a single mitochondrion, and the suggestion is that we don't.

Shirihai: It is in a paper that we published (Twig et al 2006). We looked at networks that are convoluted, using the same methodology but doing a nice z-stack to look at this possibility. I would add that 5 mV change in potential means a large change in fluorescence intensity of TMRE. Given that the changes we see in mitochondria during the period in which it does not go through fusion or fission are 1–2 mV, the changes observed under fission cannot be considered as noise.

Halestrap: What is the resolution in the z dimension of a confocal laser microscope compared with the thickness of the mitochondrion?

Shirihai: The xy or planar resolution is dependent on the wavelength and the numerical aperture and is limited to 0.2 μm. The z axis resolution is limited by the density of the z stack slices as set for the experiment and in most systems is limited to 0.5 μm. z axis resolution can be improved by deconvolution and by the use of a 2-photon laser for excitation. In theory, z axis resolution can reach 0.2 μm.

Halestrap: The narrowest bit of a mitochondrion would be around 1 μm.

Lemasters: Let's not forget that there is more to the protonmotive force than membrane potential. The loss of membrane potential might indicate an increase in pH gradient, in which case oligomycin effects would not be much changed. It is also possible for volume changes to occur such as those Charles Hackenbrock described: the cristae expand, and matrix volume becomes smaller (Hackenbrock 1966, 1968). Consequently, less dye is taken up even if membrane potential is unchanged. Cristae are below the resolving power of optical microscopes, and thus such changes may escape detection.

Nicholls: One test that could be done is what we call the 'oligomycin null point test'. If oligomycin is added, healthy mitochondria hyperpolarize. A mitochondrion that is reversing its ATP synthase because it has a damaged respiratory chain will depolarize. It would be interesting to confirm with your depolarized mitochondria that the membrane potential changes in the direction you would predict.

Shirihai: We tested the oligomycin reversal point analysis. We saw that the difference between the daughter mitochondria stays the same or increases. We added oligomycin during the fission event at an early stage. The mitochondrion that was more polarized became even more polarized, and the one that is low stays low. However, due to the fact that after fission mitochondrial membrane potential is changing, the actual effect of oligomycin on the membrane potential of the depolarized daughter unit in millivolts could not be calculated with accuracy.

Schon: If this were a patient cell and it was heteroplasmic, by Muller's ratchet you would eventually purify the cell of the mutation. Muller's ratchet is the biological version of Gresham's law, itself an economic version of Murphy's law, which is that left to their own devices, things go to hell. Gresham's law is that bad money drives out good. If you have a heteroplasmic situation with good and bad mitochondria, and the good ones are fused and normally polarized, and the bad ones are sub-polarized, you would expect that over time the heteroplasmy would go towards normality because of the autophagy. In most cases this does not happen. There is a remarkable stability of heteroplasmy. In the context of this conversation, there are two kinds of diseases: respiratory chain mutations, where heteroplasmy destroys proton pumping, and then there are ATPase mutations where the polarization is increased. In both classes of disease there is no Muller's ratchet.

Turnbull: There might be a difference between different cell types. Most studies in patients are done on post-mitotic cells. There are also some studies on dividing cells. From what we have heard I can't believe it is just based on membrane potential, from what we see in the patients and their cells. As Erich points out, otherwise you would lose these mutations. Everything is saying that we don't lose the mutations, and with ageing we accumulate mutations.

Duchen: Has anyone convincingly shown heterogeneity in mitochondrial membrane potential in single cells from patients who have heteroplasmic mitochondria?

Schon: Yes, but it has been done by JC1. Hardly anyone does TMRE.

Duchen: We have looked at fibroblasts using TMRM and haven't seen heteroplasmy. If there is fission and fusion, and sharing of proteins, then the mutant protein load would be dispersed among the population. You might not expect to find mitochondrial heterogeneity.

Rich: I'd like to explore possible molecular mechanisms for the depolarization that you see. One way that it could occur is by decreasing substrate supply although, because of the non-ohmic nature of the mitochondrial membrane, a large decrease in supply and respiratory rate would be required to produce an observable decrease of membrane potential. Such drastic substrate limitation would require some additional molecular mechanism. Or, as I understand from the talk so far, you may instead be thinking more in terms of there being some kind of membrane damage which increases the permeability of the membrane in some way. In this case, the depolarized mitochondria would instead be partly uncoupled and respiring much faster. Is this how people think? In this latter case there would be a population of mitochondria that are significantly uncoupled and running faster and their effect on the efficiency of the cell could be dramatic and possibly detectable, depending on the proportion of mitochondria that they represent.

Bernardi: If there is an increase in permeability there might be a release of pyridine nucleotides, and the mitochondria would end up respiring slower not faster, even though the inner membrane permeability is higher.

Halestrap: The oligomycin didn't put up the potential, which you'd expect for healthy mitochondria.

Rich: Has anyone, for example, tried to measure the rates of respiration in these two types of mitochondria? I'm not sure how you'd do it, but it could possibly be done by looking at local oxygen concentrations with an oxygen probe. Mitochondria running faster would have a lower local oxygen concentration.

Larsson: Autophagocytosis is involved in clearance of defective mitochondria. In your system you have co-localization with markers for autophagocytosis and defective mitochondria. I have also seen published EMs showing that autophagocytosis may occur *in vivo* (Tyynismaa et al 1995). Presumably there are mouse mutants with defective autophagocytosis: do they also have a problem with the clearance of defective mitochondria? This would provide genetic evidence for a role for autophagocytosis in the clearance of mitochondria.

Shirihai: We have looked at the ATG5 knockout mouse which lacks autophagy. We see here an accumulation of mitochondria with increased heterogeneity in membrane potential.

Lemasters: We should not think that all autophagy is linked to mitochondrial depolarization. I think it will be more complicated. But if mitochondria are damaged, they seem to be targeted for autophagy. We have done this using a laser to selectively photodamage a small group of mitochondria (Kim et al 2007). After 25 minutes we see GFP-LC3-labelled rings form around these mitochondria. LC3 is a specific autophagy marker. Thus, mitochondrial damage seems to be a sufficient signal for autophagy. But whether there are instances where autophagy is initiated by other means remains to be seen. In nutrient deprivation, depolarization of mitochondria actually occurs as the autophagosome forms, and isolation membranes form well before depolarization of the inner membrane. By contrast, if we just zap individual mitochondria and wait, the damaged and depolarized individual mitochondria will be engulfed and eliminated.

Nicholls: Do we have an idea of how a mitochondrion that will be targeted for autophagocytosis is recognized?

Lemasters: That's the big question.

Halestrap: Have people done the experiment with oligomycin and uncouplers, so that there are uncoupled mitochondria but the cells are maintaining their ATP levels? Is there an increase in mitochondrial autophagy?

Lemasters: I haven't looked at this.

Halestrap: This would quickly tell you whether depolarization is key.

Shirihai: Oligomycin by itself is initiating autophagy, but this could be due to something similar to nutrient deprivation. A brief FCCP window that can be recovered later does not initiate massive autophagy.

Halestrap: It is the two together: having the uncoupler to depolarize and the oligomycin to prevent the mitochondrial ATPase from hydrolysing ATP derived from glycolysis.

Bernardi: It would kill the cell rapidly.

Scorrano: We have tried to see for ourselves whether fission could be a signal for targeting mitochondria to autophagy. We have some preliminary data suggesting that we can separate targeting to the autophagosome from depolarization. We can induce fission without depolarizing mitochondria, yet target them to the autophagosome.

Hajnóczky: Orian Shirihai showed nicely that mitochondrial fusion occurs frequently in β cells. Practically, the entire mitochondrial population is involved in fusion in a couple of hours. He also showed that fusion is a rapidly reversed event most of the time. Most of the mitochondria that undergo this cycle stay as good as they look. Perhaps we should consider the possibility that this fusion–fission cycle and the ensuing mixing of some mitochondrial factors in the entire mitochondrial population is a component of maintaining the functional stability of the mitochondria.

Another aspect of the fusion dynamics that has not been brought up so far is that fission is necessary for producing discrete structures that can be targeted throughout the cell, whereas fusion by itself would produce a continuous network that has limited mobility and is probably confined to a smaller cellular region. It is important to consider fission as a part of the physiological maintenance of the mitochondrial mobility in this case.

Rizzuto: Do I recall correctly that mitochondrial myopathies maintain segmental defects across the fibre? In a way, this contradicts what has been elegantly shown in terms of rapid diffusion.

Schon: Muscle is a problem for a couple of reasons. First, it is a syncitium. Second, if this room were a muscle you wouldn't be able to move from one side to another:

it is so packed with proteins. The timescale is months and years, as opposed to minutes and hours. Muscle is a hard place to look for the phenomenon. In other cells, the typical nature of mitochondria harbouring pathogenic mtDNA mutations is punctate.

Scorrano: Can it be that in muscle, the mitochondrial unit that carries this level of mtDNA mutation is unable to support protein import as efficiently as the neighbouring mitochondrion? Therefore it would be unable to fuse because it lacks the fusion components, which are encoded by the nucleus.

Schon: No. It is capable of importing all kinds of things.

Scorrano: No one looks at individual mitochondria: we always look at a population.

Schon: You can look in a muscle fibre that is completely knocked out for respiratory chain function and it can import succinate dehydrogenase.

Jacobs: We are skirting round an issue that we haven't really addressed, and which your experiments say we need to address, which is 'what is happening to DNA during these fusion/fission events?' What is happening to nucleoids? Are mitochondria that become dark those that have lost nucleoids completely, or where there are 'bad' nucleoids? You surely have the tools to answer this question, namely photoactivatable GFP which could be targeted to nucleoids.

References

Bach D, Pich S, Soriano FX et al 2003 Mitofusin-2 determines mitochondrial network architecture and mitochondrial metabolism. A novel regulatory mechanism altered in obesity. Biol Chem 278:17190–17197

Chen H, Chomyn A, Chan DC 2005 Disruption of fusion results in mitochondrial heterogeneity and dysfunction. J Biol Chem 280:26185–26192

Hackenbrock CR 1966 Ultrastructural bases for metabolically linked mechanical activity in mitochondria. I. Reversible ultrastructural changes with change in metabolic steady state in isolated liver mitochondria. J Cell Biol 30:269–297

Hackenbrock CR 1968 Chemical and physical fixation of isolated mitochondria in low-energy and high-energy states. Proc Natl Acad Sci USA 61:598–605

Kim I, Rodriguez-Enriquez S, Lemasters JJ 2007 Selective degradation of mitochondria by mitophagy. Arch Biochem Biophys 462:245–253

Lemasters JJ, Ramshesh VK 2007 Imaging of mitochondrial polarization and depolarization with cationic fluorophores. Methods Cell Biol 80:283–295

Malka F, Guillery O, Cifuentes-Diaz C et al 2005 Separate fusion of outer and inner mitochondrial membranes. EMBO Rep 6:853–859

Overly CC, Rieff HI, Hollenbeck PJ 1996 Organelle motility and metabolism in axons vs dendrites of cultured hippocampal neurons. J Cell Sci 109:971–980

Scorrano L, Petronilli V, Colonna R, Di Lisa F, Bernardi P 1999 Chloromethyltetramethylrosamine (Mitotracker Orange) induces the mitochondrial permeability transition and inhibits respiratory complex I. Implications for the mechanism of cytochrome c release. J Biol Chem 274:24657–24663

Smiley ST, Reers M, Mottola-Hartshorn C et al 1991 Intracellular heterogeneity in mitochondrial membrane potentials revealed by a J-aggregate-forming lipophilic cation JC-1. Proc Natl Acad Sci USA 88:3671–3675

Twig G, Graf SA, Wikstrom JD et al 2006 Tagging and tracking individual networks within a complex mitochondrial web with photoactivatable GFP. Am J Physiol Cell Physiol 291: C176–C184

Tyynismaa H, Mjosund KP, Wanrooij S et al 2005 Mutant mitochondrial helicase Twinkle causes multiple mtDNA deletions and a late-onset mitochondrial disease in mice. Proc Natl Acad Sci USA 102:17687–17692

Multiple functions of mitochondria-shaping proteins

Luca Scorrano

Dulbecco-Telethon Institute, Venetian Institute of Molecular Medicine, Via Orus 2, 35129 Padova, Italy

Abstract. Mitochondria are complex organelles whose internal structure and cytosolic organization is controlled by a growing number of 'mitochondria-shaping' proteins. These include mitochondrial proteins such as the large dynamin-related GTPases Mitofusin (Mfn) 1 and 2, Optic Atrophy 1 (Opa1); as well as the cytosolic dynamin-related protein 1 (Drp1) and its receptor on the outer mitochondrial membrane Fis1. These proteins influence not only the shape of mitochondria, but also the function of the organelle and eventually integrated cellular signalling cascades, including apoptosis. We undertook a genetic approach to elucidate the function and regulation of these proteins. Opa1 is involved in the regulation of mitochondrial fusion, by co-operating with Mfn1. Moreover, Opa1 independently from mitochondrial fusion regulates the crista remodelling pathway of apoptosis. Oligomers of a membrane bound and a soluble form of Opa1, produced by Parl, an inner membrane rhomboid protease, are disrupted early during apoptosis, leading to remodelling of the mitochondrial cristae and redistribution of the mitochondrial cytochrome c. The importance of this pathway is substantiated by the phenotype of the *Parl*$^{-/-}$ mouse, which displays excess apoptosis in multiple tissues. Cells lacking Parl are more susceptible to apoptotic stimuli and the reintroduction of a soluble form of Opa1 rescues their phenotype.

2007 Mitochondrial biology: new perspectives. Wiley, Chichester (Novartis Foundation Symposium 287) p 47–59

Mitochondria are versatile organelles. Not only do they provide most ATP, but they also participate as key players in several signalling pathways crucial for the life and death of the cell. They shape cytosolic Ca^{2+} transients by taking up Ca^{2+} ions following release from the endoplasmic reticulum stores (Rizzuto et al 1992); they integrate apoptotic signals by releasing proapoptotic factors that are required to activate the caspases involved in the dismissal of the constituents of the cell (Danial & Korsmeyer 2004); and they are part of essential anabolic and catabolic pathways that generate or dispose cellular constituents or molecules involved in signalling pathways (Ernster & Schatz 1981). In sum, they are fulfilling a plethora of additional functions beside their role in ATP biogenesis. To play these multiple roles, mitochondria adopt complex shapes, finely regulated by a specific machinery of fusion and fission (Dimmer & Scorrano 2006).

Mitochondrial shape is extremely complex, with the organelle bound by two membranes. Moreover, the inner membrane is further subdivided in two distinct compartments, the peripheral inner boundary membrane and the cristae (Frey & Mannella 2000). As shown by several lines of evidence, cristae represent the site of oxidative phosphorylation, while the inner boundary membrane is enriched in structural proteins and in components of the import machinery of mitochondria (Vogel et al 2006). When we analyse the morphology of the organelle in the cytosolic context we observe another level of complexity, given by their organization as single, separated mitochondria, or in networks of interconnected organelles. It is clear that this structural complexity results from the integration of multiple layers of control. First, a machinery that regulates fusion and fission of individual organelles should be in place (Dimmer & Scorrano 2006); second, the relative position of mitochondria has to be regulated by the interaction with the scaffolds of the cell: microtubules, thick filaments and intermediate filaments (Anesti & Scorrano 2006). Third, the complexity of the inner membrane should not be left to the biophysics of the different lipids that characterize it, like cardiolipin, but is likely to depend on a specific set of regulatory proteins. Finally, all these controllers are to be integrated to meet the specific requirements of the cell and to respond to physiological and pathological stimulation.

Our knowledge of the proteins that regulate mitochondrial shape, or 'mitochondria-shaping' proteins, is continuously increasing. Pioneering work in the yeast *Saccharomyces cerevisiae* identified most of the players in this pathway (Shaw & Nunnari 2002). Homologues for some but not all proteins of this family have been identified in higher eukaryotes (Karbowski & Youle 2003). Whether the mammalian machinery recapitulates the yeast one is a matter of debate. In particular, mammalian mitochondria participate, as outlined above, in several pathways other than the ones typical of yeast.

It is therefore unclear whether mammalian cells evolved a specific set of mitochondria-shaping proteins to fulfil their specific needs, or whether the basic sets of these proteins acquired additional functions. Nevertheless, the importance of these proteins in mammals is substantiated by at least two genetic diseases: Charcot-Marie-Tooth IIa (Zuchner et al 2004) and dominant optic atrophy (Alexander et al 2000, Delettre et al 2000).

In the following pages we will summarize our recent work aimed at characterizing the functional complexity of mammalian mitochondria-shaping proteins. In particular, we will focus our attention on Opa1 and its multiple roles in apoptosis and mitochondrial shape; and on the functional difference between Mitofusin 1 and 2. Experimental details on the data presented here can be found elsewhere (Cipolat et al 2004, 2006, Frezza et al 2006).

Results and discussion

The release of cytochrome c during apoptosis is remarkably complete and usually occurs without swelling and gross rupture of the mitochondrial architecture. This poses the problem of the mechanism by which the protein, mostly located at the level of the cristae, can be released. Back in 2002 we described a pathway of 'remodelling' of the mitochondrial cristae during apoptosis, characterized by the fusion of individual cristae and more importantly by the opening of the narrow tubular cristae junction. This results in mobilization of the cristae cytochrome c stores towards the intermembrane space, from where it can then be released across the outer membrane (Scorrano et al 2002). The molecular mechanisms governing the shape of the cristae as well as their remodelling during apoptosis are largely unknown, but a natural candidate to influence these processes is Opa1, an inner mitochondrial membrane dynamin-related protein mutated in dominant optic atrophy (Delettre et al 2000).

Opa1 is crucially involved in mitochondrial fusion: levels of Opa1 are directly proportional to the extent of fusion, measured in polyethylene glycol fusion assays, and therefore to elongation of the organelle (Fig. 1). Moreover, Opa1 requires Mitofusin 1 but not Mitofusin 2 to drive fusion of the organelles (Cipolat et al 2004). Given the ability of several pro-fusion mitochondrial proteins to influence apoptosis, it was therefore essential to measure whether Opa1 had any effect on death.

Expression of Opa1, but not of mutants resembling the clinically relevant mutations, in mouse embryonic fibroblasts (MEFs) protected from death by intrinsic, mitochondria utilizing stimuli. This protection occurred at the mitochondrial level, as substantiated by the reduced cytochrome c release and mitochondrial

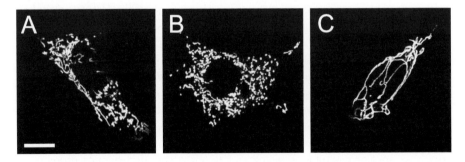

FIG. 1. Mitochondrial morphology is controlled by Opa1. MEFs cotransfected with mitochondrially targeted YFP (mtYFP) (A), with mtYFP and a vector coding for shRNA against Opa1 as described in Cipolat et al (2004) (B), or with mtYFP and Opa1 (C) were imaged after 24 h. Representative confocal images show different appearance of the mitochondrial network. Bar, 15 μm.

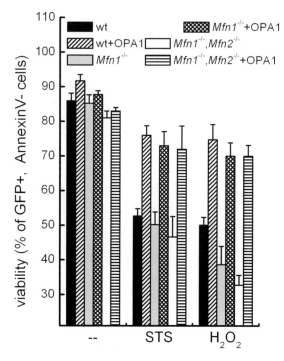

FIG. 2. Opa1 protects from apoptosis independently of mitofusins. MEFs of the indicated genotype were cotransfected with GFP and the indicated plasmids. After 24 h, cells were treated for 6 h with staurosporine (STS, 2 μM) or H_2O_2 (1 mM) and viability was determined cytofluorimetrically as described in Frezza et al (2006).

dysfunction. Moreover, a genetic approach of expressing Opa1 in MEFs deficient for Mitofusin 1 or both mitofusins revealed that the protection was independent from the pro-fusion effect of Opa1 (Fig. 2). Therefore Opa1 protects from apoptosis when it cannot shift the fusion–fission equilibrium towards fusion.

To further investigate the mechanism of action of Opa1 we generated a set of cellular models of Opa1 up- and down-regulation. The mitochondria isolated from these models were then treated with recombinant Bid and the release of cytochrome c was measured semi-quantitatively by using an ELISA assay. Moreover, this approach allowed us to measure the redistribution of cytochrome c from the cristae to the intermembrane space and to visualize the internal structure of normal and apoptotic mitochondria by electron tomography. To summarize the results obtained by this approach, we could identify an inverse relationship between levels of Opa1 and cristae remodelling in response to Bid. Mitochondria overexpressing Opa1 were markedly resistant to mobilization of cytochrome c and

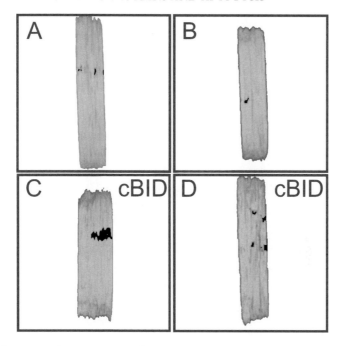

FIG. 3. Representative tomograms of normal and apoptotic mitochondria. Mitochondria isolated from wild-type (A, C) and Opa1-overexpressing (B, D) MEFs (Frezza et al 2006) were treated where indicated with 30 pmol/mg caspase-8 cleaved BID for 5 min, fixed and processed for electron tomography as described in Frezza et al (2006). Following reconstruction and volume rendering, tomograms were computer processed to peel the outer membrane and visualize only the inner membrane (grey) and the cristae opening (black; background is in white). Note that in Opa1-overexpressing mitochondria BID is unable to enlarge the cristae junction. Details on the tomographic procedure are in Frezza et al (2006).

remodelling of the cristae (Fig. 3), whereas down-regulation of Opa1 by RNA interference, or expression of a mutant Opa1, precipitated both responses to Bid.

The mechanism by which Opa1 could control the shape of the cristae and more specifically the diameter of the cristae junction is unclear. Dynamins usually form rings around the biological membranes they tubulate or sever (McNiven 1998). On the other hand, Opa1 is located inside of the membrane of the cristae and therefore an alternative mechanism should be operational. When we analysed the submitochondrial distribution of Opa1 we found a small fraction of the protein soluble in the intermembrane space. We reasoned that Opa1 could form oligomers inside the cristae, comprising both the membrane bound and the soluble form. A combination of size exclusion chromatography and chemical cross-linking confirmed that this was the case, showing Opa1 oligomers that disappeared following mechanical distension of the cristae. Of note, these oligomers were early targets during Bid-induced remodelling of the cristae and were stabilized by the over expression of Opa1.

How is the soluble form of Opa1 generated? It is well known that multiple splice variants of Opa1 exist, raising the possibility that this minor soluble form was the result of alternative splicing. Conversely, in yeast the orthologue of Opa1 is post-translationally cleaved by a specific inner mitochondrial membrane rhomboid protease (McQuibban et al 2003). Rhomboids are intramembrane proteases that cleave transmembrane proteins in the context of the membrane (Urban et al 2001). The mammalian homologue of this mitochondrial rhomboid is Parl. In order to address the role of Parl in this pathway, we performed a genetic analysis. Mutant mice lacking *Parl* display normal intra-uterine development and are normal at birth and up to 4 weeks. Then they stop growing, lose weight and enter a progressive cachectic status sustained by atrophy of multiple organs such as the thymus, spleen and skeletal muscle. This generalized wasting syndrome leads to death between weeks 8 and 12. *In situ* and *ex vivo* analysis of the affected tissues showed that atrophy was sustained by increased apoptosis and by an increased response of $Parl^{-/-}$ cells to apoptotic stimuli. We then turned to $Parl^{-/-}$ MEFs and verified whether Opa1 and Parl lie in the same genetic pathway of apoptosis. Expression of Opa1 in $Parl^{-/-}$ MEFs did not rescue from death by intrinsic stimuli. Moreover, reintroduction of Parl in $Parl^{-/-}$ MEFs where expression of Opa1 was silenced similarly did not complement the apoptotic defect of these cells, showing that Parl is upstream of Opa1 in this death pathway.

The relative amount of soluble Opa1 was reduced in $Parl^{-/-}$ tissues by approximately 50%. The reintroduction of wild-type, but not of catalytically inactive Parl in $Parl^{-/-}$ MEFs corrected the levels of soluble Opa1. The reduction in soluble Opa1 levels correlated with increased remodelling of the cristae, faster disruption of Opa1 oligomers, higher mobilization and release of cytochrome c and ultimately with greater apoptosis (Fig. 4). A mutant of Opa1 targeted to the intermembrane space corrected the defect of $Parl^{-/-}$ MEFs, further substantiating the importance of this soluble form of Opa1 in the control of apoptosis.

Conclusions

Our data outline the multiple functions of Opa1, a mitochondria-shaping protein. On the one hand it participates in the control of mitochondrial morphology, by impinging on fusion of the organelles (Cipolat et al 2004). On the other, it controls apoptosis independently of fusion, by regulating the cristae remodelling pathway (Frezza et al 2006). Parl, the only mitochondrial rhomboid protease identified so far, is at the crossroads between these two functions of Opa1. While its ablation apparently does not affect mitochondrial shape, it impacts on the regulation of

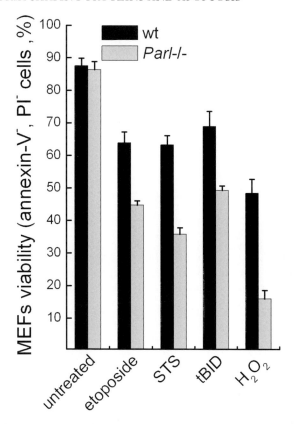

FIG. 4. $Parl^{-/-}$ MEFs are more susceptible to apoptosis by intrinsic stimuli. MEFs of the indicated genotype were treated for 6 h with staurosporine (STS, 2 μM) or H_2O_2 (1 mM) or for 48 h with etoposide (2 μM) and viability determined as described in Cipolat et al (2006). Where indicated, cells were cotransfected with GFP and tBID and viability was determined after 24 h.

apoptosis and in particular on the pathway controlled by Opa1 (Cipolat et al 2006). The complex phenotype of the $Parl^{-/-}$ mouse suggests that the crista remodelling pathway, in which Parl crucially participates, is important for tissue homeostasis in the adult animal. Mice lacking pivotal apoptosis regulators, such as Bcl-2, display only limited developmental problems and a phenotype closely similar to that of $Parl^{-/-}$ mice (Veis et al 1993). It is tempting to speculate that mitochondrial pathways of apoptosis are mostly recruited for programmed death of adult tissues, rather than for development; and that the cristae remodelling pathway is a core component of mitochondrial death.

Whether Opa1 is directly cleaved by Parl remains an open question. The other inner membrane protease paraplegin has been suggested to play a role in the processing of Opa1 (Ishihara et al 2006). In pulse-chase experiments of radioactive labelled Opa1 presented by Mihara and colleagues, processing of Opa1 is slowed down when Parl levels are reduced and enhanced by silencing of paraplegin. On the other hand, paraplegin seems to affect mostly the pro-fusion function of Opa1, independent of Parl (Cipolat et al 2004). It is possible that both proteases can process Opa1. It remains to be seen whether they influence different functions of this protein. This will be a challenging task for the future of our research.

References

Alexander C, Votruba M, Pesch UE et al 2000 OPA1, encoding a dynamin-related GTPase, is mutated in autosomal dominant optic atrophy linked to chromosome 3q28. Nat Genet 26:211–215
Anesti V, Scorrano L 2006 The relationship between mitochondrial shape and function and the cytoskeleton. Biochim Biophys Acta 1757:692–699
Cipolat S, de Brito OM, Dal Zilio B et al 2004 OPA1 requires mitofusin 1 to promote mitochondrial fusion. Proc Natl Acad Sci USA 101:15927–15932
Cipolat S, Rudka T, Hartmann D et al 2006 Mitochondrial rhomboid PARL regulates cytochrome c release during apoptosis via OPA1-dependent cristae remodeling. Cell 126:163–175
Danial NN, Korsmeyer SJ 2004 Cell death: critical control points. Cell 116:205–219
Delettre C, Lenaers G, Griffoin JM et al 2000 Nuclear gene OPA1, encoding a mitochondrial dynamin-related protein, is mutated in dominant optic atrophy. Nat Genet 26:207–210
Dimmer KS, Scorrano L 2006 (De)constructing mitochondria: what for? Physiology (Bethesda) 21:233–241
Ernster L, Schatz G 1981 Mitochondria: a historical review. J Cell Biol 91:227s–255s
Frey TG, Mannella CA 2000 The internal structure of mitochondria. Trends Biochem Sci 25:319–324
Frezza C, Cipolat S, Martins DB et al 2006 OPA1 Controls apoptotic cristae remodeling independently from mitochondrial fusion. Cell 126:177–189
Ishihara N, Fujita Y, Oka T et al 2006 Regulation of mitochondrial morphology through proteolytic cleavage of OPA1. EMBO J 25:2966–2977
Karbowski M, Youle RJ 2003 Dynamics of mitochondrial morphology in healthy cells and during apoptosis. Cell Death Differ 10:870–880
McNiven MA 1998 Dynamin: a molecular motor with pinchase action. Cell 94:151–154
McQuibban GA, Saurya S, Freeman M 2003 Mitochondrial membrane remodelling regulated by a conserved rhomboid protease. Nature 423:537–541
Rizzuto R, Simpson AW, Brini M et al 1992 Rapid changes of mitochondrial Ca2+ revealed by specifically targeted recombinant aequorin. Nature 358:325–327
Scorrano L, Ashiya M, Buttle K et al 2002 A distinct pathway remodels mitochondrial cristae and mobilizes cytochrome c during apoptosis. Dev Cell 2:55–67
Shaw JM, Nunnari J 2002 Mitochondrial dynamics and division in budding yeast. Trends Cell Biol 12:178–184
Urban S, Lee JR, Freeman M 2001 Drosophila rhomboid-1 defines a family of putative intramembrane serine proteases. Cell 107:173–182
Veis DJ, Sorenson CM, Shutter JR et al 1993 Bcl-2-deficient mice demonstrate fulminant lymphoid apoptosis, polycystic kidneys, and hypopigmented hair. Cell 75:229–240
Vogel F, Bornhovd C, Neupert W et al 2006 Dynamic subcompartmentalization of the mitochondrial inner membrane. J Cell Biol 175:237–247

Zuchner S, Mersiyanova IV, Muglia M et al 2004 Mutations in the mitochondrial GTPase mitofusin 2 cause Charcot-Marie-Tooth neuropathy type 2A. Nat Genet 36:449–451

DISCUSSION

Speigelman: Given the history of how the function of Bcl-2 was discovered by your former mentor, you might wonder about activating mutations in Opa1 in human cancer.

Scorrano: We are doing these experiments. We thought if we could at least delay apoptosis this could play a role in sensitizing or accelerating tumorigenesis. We prepared an Opa1 transgenic mouse under the control of a generic promoter. We are at F3, so we are still cleaning the background of these animals. Surprisingly, the animals are bigger then their littermates. They aren't fatter; just 30% bigger. We are now trying to exclude any contribution from the background. When we analyse histology we see that the cells are actually bigger.

Orrenius: Going back to the Opa1 effect, if you look at other proteins that are released from the mitochondria in response to apoptotic stimuli, would you see a similar response with those? I'm referring to proteins which normally have no known association with the inner membrane.

Scorrano: Cytochrome c is kind of special because it is compartmentalized. It is mostly located in the cristae, and associated with the inner mitochondrial membrane (IMM). I'm not sure whether other proteins such as Smac or AIF share the same location inside the mitochondria. For us it is easier to measure submitochondrial localization of cytochrome c because we can use the redox properties of this molecule. It is more difficult to do this with Smac because we would need to look at the electron microscopy (EM) level and then measure the number of dots in the cristae versus those in the intermembrane space, which is less quantitative.

Orrenius: Along the same lines, during Bcl-2 protein-mediated permeabilization we believe that pores open in the outer membrane, but that the inner membrane remains relatively intact. To understand the Opa1 effect better, if you were to release cytochrome c by inducing permeability transition pore (PTP) opening, would you still see the Opa1 effect?

Scorrano: Some of the stimuli which we used, such as hydrogen peroxide, cause the release of cytochrome c and are PTP inducers, and we still see that Opa1 has a slowing effect. Back in 1999 Paolo suggested that extensive flickering of the PTP could contribute to the remodelling of the cristae. Perhaps this is the level at which the two pathways cross-talk. We should go into this in more detail and try to understand the cross-talk between PTP and the cristae remodelling controlled by Opa1.

Nicholls: Can we focus on the cristae remodelling because I am struggling to understand this. Many years ago Charlie Hackenbrock came up with his orthodox and condensed conformations (Hackenbrock 1966), which turned out to be secondary to ion transport mechanisms across the inner membrane. Do you have any contribution from osmotic effects in the remodelling, or can it purely be accompanied by the protein interactions?

Scorrano: During the remodelling of the cristae we see an increased flickering of the PTP, so there are probably ion fluxes across the IMM.

Nicholls: Including cyclosporin A or adenine nucleotides in the medium dramatically inhibits PTP opening. Does this stop the remodelling?

Scorrano: Yes. This is what we published back in 2002 (Scorrano et al 2002)

Halestrap: I don't know whether I'd agree that the ion fluxes are responsible for the changes from orthodox to condensed conformations. You can reproduce changes between the two states by the addition of atractyloside and bongkrekic acid that change the adenine nucleotide translocase conformation to the 'm' and 'c' state respectively (Stoner & Sirak 1973a,b). Thus it could be that it is changes in protein conformation in the inner membrane rather than matrix volume that is having effects on the cristae.

Scorrano: There are several reports of the effect of ATPase on the shape of the cristae (e.g. Gavin et al 2004). We measured ATPase complexes during cristae remodelling and could not see any change. It may be true in yeast where the shape of the cristae is different from that in mammalian mitochondria. In mammalian mitochondria the contribution of respiratory chain components is less dominant in regulating the shape of the cristae.

Lemasters: Going back to Charles Hackenbrock and also David Green who worked on energy-dependent changes in light scattering, it turns out that there are two different mechanisms contributing to the changes. One is the mitochondrion acting as an osmometer, swelling as ions are taken up and shrinking as they are released. The other has to do with changes in the actual curvature of the inner membrane. Atractyloside binds on the outside of the inner membrane and produces a predominantly convex folding from the inner membrane with consequent changes of light scattering, as shown by Stoner & Sirak (1973b). Bongkrekic acid, which binds to the inner surface of the inner membrane, does the opposite. Binding of bongkrekic acid and atractyloside cause large changes in light scattering, particularly in heart mitochondria, that are independent of changes in matrix volume. Superimposed on light scattering changes caused by changes in membrane curvature are those caused by changes in mitochondrial volume. In particular, the effects caused by bongkrekic acid and atractyloside (or carboxyatractyloside) are very much like the cristae remodelling that Luca describes. Stoner and Sirak did this in the 1970s (Stoner & Sirak 1973b).

O'Rourke: In the early studies of swelling and contraction of mitochondria there was an ATP-induced contraction described by Lehninger (1962). This was an independent process from the osmotic swelling. It was ATP dependent but not energy dependent. Is Opa1 in any way related to this contraction, which he envisaged as a motile process on the mitochondrial membrane?

Scorrano: That is a good question but we have no idea. All I can say is that we can disrupt the oligomers when we swell the mitochondria. We haven't checked whether by reversing the opening of the PTP by adding cyclosporine A we could restabilize the oligomers. They come in place only when the proteins are physically close enough.

O'Rourke: I was thinking more in the physiological domain. Have you looked at State IV to State III transitions in mitochondrial structure to see whether Opa plays a role?

Scorrano: No, but we should.

Nicholls: Your hypothesis predicts that the Opa1 oligomers are at the neck of the cristae, near the inner membrane. Can this be seen by EM?

Scorrano: We tried doing this with electron tomograms, so that we could localize them precisely. But we need better antibodies. Electron tomograms are done with thick sections, so the antibodies need to penetrate well. We are trying to produce an antibody that would allow us to visualize the mitochondrial localization of Opa1. A recent paper (Vogel et al 2006) showed that the Opa1 orthologue, Mgm1p, is mostly concentrated along the cristae with very little on the inner membrane. It seemed that there was a slight accumulation of Mgm1 at the neck of the cristae. I asked Andy Reichert whether he though there was a concentration here, and he said he wasn't sure.

Youle: Dynamin forms spirals around membranes, and Drp1 can form spirals. Opa1 is in the same family, so could it also form a spiral? There is a similar issue with the process of formation of multivesicular bodies, in which a lysosome internalizes into bodies on the inside which are pinched off. There is some dynamin/VPS involved in this.

Scorrano: That is very likely. Chemically, we can stabilize this trimer of Opa1. We did gel fractionation studies and we can find it in high molecular weight complexes. I don't know whether they are formed only of Opa1 or whether there are accessory and regulatory proteins in this. It is likely that they multimerize.

Rizzuto: I'd like to go back to a different topic: that of Fis1. These are nice data because they tell us we are bound to complexity. We all know that most apoptotic cells have fragmented mitochondria, and this led to the association of fission with apoptosis. This appears a simplification. Your data show that apoptosis is unrelated to the fission activity of Fis1, and Drp1 overexpression doesn't kill cells. Rather, we know that fragmentation protects against some apoptotic stimuli. We have to live with the fact that of course we have to make a simplified scheme in order to understand what is going on, but we should recognize that simplifications generate misconceptions. In the real world proteins can have different effects. For the organelle or cytochrome c this is not a surprise. One day we will have the full picture, but for the time being we shouldn't be making the easy connection 'fission = apoptosis'. This will mislead a number of people.

Youle: I don't know anyone who has proposed that fission = apoptosis. You can obviously generate fission without inducing apoptosis. The more interesting question is can you get apoptosis in the absence of fission?

Bernardi: I have a comment on the fact that in my view this direct connection between pro-fission proteins and apoptosis may provide a final common pathway. Otherwise there is a paradox. Opa1 apparently is able to do this job irrespective of the cell type it is expressed in. And yet, when it is missing, the only effect is optic nerve atrophy. Either there is something in other cells that does the same job (it could be Fis1), or else we have a problem because you showed in MEFs that you can get the same result in any cell type.

Nicholls: How do you see the relationship between the PTP and the remodelling?

Bernardi: I have never studied fusion and fission proteins, but cristae remodelling can cause increased availability of cytochrome c for release. You can do this in any isolated mitochondrion: pore opening will somehow remodel the cristae before the outer membrane breaks. We never addressed the question of the proteins that are involved, but you could also have remodelling that precedes and favours pore opening.

Scorrano: Our knowledge will be helped by knowing the molecular nature of the PTP!

Bernardi: One extra issue. Let's not label cyclosporin as a blocker of the pore.

Martinou: If I understand correctly, the phenotype of the *Parl1* knockout is due to excess apoptosis occurring after birth. This suggests that following stress, the cells would release some cytochrome c. However in normal cells this level would not be sufficient to trigger apoptosis. In contrast, in cells lacking Parl1 a high level of cytochrome c would be released leading to cell death. Would this be your explanation?

Scorrano: Yes, this is a possible explanation.

Martinou: How do you explain that during development, when there is massive cell death occurring, you do not get excess apoptosis?

Scorrano: The mitochondrial pathway itself doesn't play an important role in developmental cell death. Bcl-2 knockout mice have a phenotype similar to that of Parl knockouts. The mice are normal at birth and don't have developmental defects.

Adam-Vizi: Is there any relationship between fission and reactive oxygen species (ROS) generation? A recent paper reported that hyperglycaemia induced fragmentation and ROS production (Yu et al 2006). ROS production could be prevented by blocking pyruvate entry into mitochondria and fission still occurred. But when fission was prevented there was no ROS production.

Scorrano: They didn't measure whether pyruvate was still entering at the same rate when they blocked fission. We have one experiment with *N*-acetylcysteine which prevents death by Fis1. I am always sceptical about these experiments because I can't measure ROS directly.

References

Gavin PD, Prescott M, Luff SE, Devenish RJ 2004 Cross-linking ATP synthase complexes in vivo eliminates mitochondrial cristae. J Cell Sci 117:2333–2343

Hackenbrock CR 1966 Ultrastructural bases for metabolically linked mechanical activity in mitochondria. I. Reversible ultrastructural changes with change in metabolic steady state in isolated liver mitochondria. J Cell Biol 30:269–297

Lehninger AL 1962. Water uptake and extrusion by mitochondria in relation to oxidative phosphorylation. Physiol Rev 42:467–517

Scorrano L, Ashiya M, Buttle K et al 2002 A distinct pathway remodels mitochondrial cristae and mobilizes cytochrome c during apoptosis. Dev Cell 2:55–67

Stoner CD, Sirak HD 1973a Adenine nucleotide-induced contraction of the inner mitochondrial membrane. I. General characterization. J Cell Biol 56:51–64

Stoner CD, Sirak HD 1973b Adenine nucleotide-induced contraction on the inner mitochondrial membrane. II. Effect of bongkrekic acid. J Cell Biol 56:65–73

Vogel F, Bornhovd C, Neupert W, Reichert AS 2006 Dynamic subcompartmentalization of the mitochondrial inner membrane. J Cell Biol 175:237–247

Yu T, Robotham JL, Yoon Y 2006 Increased production of reactive oxygen species in hyperglycemic conditions requires dynamic change of mitochondrial morphology. Proc Natl Acad Sci USA 103:2653–2658

Transcriptional control of mitochondrial energy metabolism through the PGC1 coactivators

Bruce M. Spiegelman

Dana-Farber Cancer Institute and the Department of Cell Biology, Harvard Medical School, Boston, MA 02115, USA

Abstract. The PGC1 transcriptional coactivators are major regulators of several crucial aspects of energy metabolism. PGC1α controls many aspects of oxidative metabolism, including mitochondrial biogenesis and respiration through the coactivation of many nuclear receptors, and factors outside the nuclear receptor family. ERRα, NRF1 and NRF2 are key targets of the PGC1s in mitochondrial biogenesis. We have recently addressed the question of the role of PGC1 coactivators in the metabolism of reactive oxygen species (ROS). We now show that PGC1α and β are induced when cells are given an oxidative stressor, H_2O_2. In fact, experiments with either genetic knockouts or RNAi for the PGC1s show that the ability of ROS to induce a ROS scavenging programme depends entirely on the PGC1s. This includes genes encoding mitochondrial proteins like SOD2, but also includes cytoplasmic proteins such as catalase and GPX1. Cells lacking PGC1α are hypersensitive to death from oxidative stress caused by H_2O_2 or paraquat. Mice deficient in PGC1α get excessive neurodegeneration when given kainic acid-induced seizures or MPTP, which causes Parkinsonism. These data show that the PGC1s are key modulators of mitochondrial biology and important protective molecules against ROS generation and damage. The implications of this for diabetes and neurodegenerative diseases are discussed.

2007 Mitochondrial biology: new perspectives. Wiley, Chichester (Novartis Foundation Symposium 287) p 60–69

Oxidative metabolism is crucial for most living systems, and the majority of oxidative metabolism in eukaryotic cells occurs in the mitochondria. Mitochondria generate ATP through the function of the electron transport system, whereby the passage of high energy electrons down this chain is coupled to the extrusion of protons across the inner membrane of the mitochondria. This proton gradient can be dissipated by passage through complex V of the electron transport chain, which couples this proton movement to the phosphorylation of ADP to ATP.

Recent data have implicated mitochondrial dysfunction in a large number of important human diseases, including neurodegeneration, heart failure and diabetes. The skeletal muscle of humans with type 2 diabetes, glucose intolerance or a family history of diabetes all have a reduced expression of multiple genes of the mitochondrial OXPHOS system, and their dominant regulators, the PGC1 (peroxisome proliferator-activated receptor γ [PPARγ] coactivator 1) coactivators (Mootha et al 2003, Patti et al 2003). Likewise, a variety of neurodegenerative disorders have been associated with dysfunctional mitochondria. In this paper, I review recent data related to the role of PGC1s, especially PGC1α, in energy metabolism related to diabetes and neurodegeneration.

PGC1α (peroxisome proliferator-activated receptor γ [PPARγ] coactivator 1α) was discovered as a binding partner and coactivator of PPARγ in brown fat (Puigserver et al 1998). It is induced in this tissue by exposure of animals to cold, a condition which activates the thermogenic function of brown fat tissue. Functional studies in our group and others showed that PGC1α increased mitochondrial OXPHOS gene expression and the expression of UCP1 when expressed in white fat cells in culture or *in vivo* (Puigserver et al 1998, Tiraby et al 2003). Detailed analysis of the effects of PGC1α on mitochondria indicated that it could increase the expression of a wide variety of mitochondrial genes, whether encoded in the nuclear or mitochondrial genomes (Wu et al 1999). This ability to activate a broad mitochondrial programme results, in large measure, from the ability of PGC1α to coactivate ERRα, NRF1 and NRF2 (Handschin et al 2003, Wu et al 1999, Mootha et al 2004, Schreiber et al 2004), and a rise in the levels of TFAM.

The ability of PGC1α and its closest homologue PGC1β to induce respiration in muscle cells was examined (St. Pierre et al 2003). Both coactivators increase respiration greatly, as they induce mitochondrial biogenesis. However, PGC1α increases the fraction of uncoupled respiration compared to controls, wheras PGC1β-induced respiration has the same relative proportions of coupled and uncoupled respiration as cells expressing a GFP control.

Regarding the role of PGC1α in skeletal muscle, the gene sets activated are not restricted to mitochondria. Transgenic expression of PGC1α stimulates a broad programme of fibre-type switching from type IIb fibres to type IIa and type I. These oxidative fibres include more mitochondria but also include myosin heavy chain (MHC) type IIa, I and myoglobin. These data are likely to be highly relevant from a physiological perspective because PGC1α is expressed at highest levels in soleus muscle, which is very rich in type I fibres (Lin et al 2002) and is induced in rodents and humans by exercise (reviewed in Handschin & Spiegelman 2006).

PGC1s and disease

Our recent studies have focused on the potential role of PGC1α in the context of tissue degeneration and wasting, especially skeletal muscle and the brain.

Since PGC1α mediates many of the effects of motor nerves on skeletal muscle relating to mitochondria and fibre-type switching, we have asked whether PGC1α might mediate a key function of motor nerve activity: the suppression of muscle atrophy. Indeed, if skeletal muscle is denervated, skeletal muscle loses mass coincident with shrinkage in the diameter of muscle fibres. This is also seen in rodents and humans if limbs suffer disuse. This loss of muscle mass is a catabolic process and is associated with induction of a set of genes termed 'atrogenes' that include E3 ubiquitin ligases such as atrogin and MURF (Lecker et al 2004).

We were able to show that dissection of the sciatic nerve causes a loss of muscle fibre diameter of 50% within 12 days in control mice. However, transgenic expression of PGC1α causes an almost complete suppression of muscle atrophy, as well as a significant reduction of the induction of the set of atrogenes (Sandri et al 2006). This suppression of the atrogenes by PGC1α is likely to derive, at least in part, from a suppression of FOXO3 action by PGC1α.

Many forms of neurodegeneration have been associated with mitochondrial dysfunction and increased oxidative damage (Lin & Beal 2006). Given that the mice mutated in PGC1α have a severe neurodegeneration in the striatum (Lin et al 2004), we recently studied the role of PGC1 coactivators and the metabolism of reactive oxygen species (ROS). PGC1α and PGC1β mRNAs are co-induced in cells treated with H_2O_2, with the gene sets of protection from ROS, including SOD1 and SOD2, catalase, GPX1, UCP1 and UCP3. Studies with RNAi directed against PGC1α show that this coactivator is required to get full expression of the antioxidant programme (St Pierre et al 2006). Similar results were obtained with cells mutated in PGC1α. The induction of PGC1α by an oxidative stressor was stimulated, at least in part, by the increased binding of phosphorylated CREB to the PGC1α promoter.

The neuroprotective and anti-ROS effects of endogenous PGC1α could be illustrated by treating the PGC1α knockout mice with agents that induce oxidative stress and neurodegeneration. MPTP (1-methyl-4-phenyl-1,2,3,6-tetrahydropyridine) and kainic acid induced much more degeneration in the dopaminergic centres of the substantia nigra, and hippocampus, respectively. These degenerative processes were associated with greater levels of stable markers of oxidative damage, such as nitrotyrosylation of proteins and 8-OXO-guanine in DNA.

Gain of function studies in cultured cells show that elevation of PGC1α levels above those of wild-type nerve cells give an increased resistance to death by the oxidative stressors H_2O_2 and paraquat.

Future studies will involve finding known drugs or chemical compounds that can elevate PGC1α in many tissues, especially brain and skeletal muscle. These will then be tested in models of neurodegeneration, muscle wasting and muscle dystrophies, and type 2 diabetes.

References

Handschin C, Spiegelman BM 2006 PGC-1 coactivators, energy homeostasis, and metabolism. Endocr Rev 27:728–735

Handschin C, Rhee J, Lin J, Tarr PT, Spiegelman BM 2003 An autoregulatory loop controls peroxisome proliferator-activated receptor γ coactivator 1α expression in muscle. Proc Natl Acad Sci USA 100:7111–7116

Lecker SH, Jagoe RT, Gilbert A et al 2004 Multiple types of skeletal muscle atrophy involve a common program of changes in gene expression. FASEB J 18:39–51

Lin J, Tarr P, Puigserver P et al 2002 Transcriptional coactivator PGC-1alpha drives the expression of slow-twitch muscle fibres. Nature 418:797–801

Lin J, Wu PH, Tarr PT et al 2004 Defects in adaptive energy metabolism with CNS-linked hyperactivity in PGC-1alpha null mice. Cell 119:121–135

Lin MT, Beal MF 2006 Mitochondrial dysfunction and oxidative stress in neurodegenerative diseases. Nature 443:787–795

Mootha VK, Lindgren CM, Eriksson KF et al 2003 PGC-1alpha-responsive genes involved in oxidative phosphorylation are coordinately downregulated in human diabetes. Nat Genet 34:267–273

Mootha VK, Handschin C, Arlow D et al 2004 Errα and Gabpβ specify PGC-1α-dependent OXPHOS gene expression that is altered in diabetic muscle. Proc Natl Acad Sci USA 101:6570–6575

Patti ME, Butte AJ, Crunkhorn S et al 2003 Coordinated reduction of genes of oxidative metabolism in humans with insulin resistance and diabetes: Potential role of PGC1 and NRF1. Proc Natl Acad Sci USA 100:8466–8471

Puigserver P, Wu Z, Park CW, Graves R, Wright M, Spiegelman BM 1998 A cold inducible coactivator of nuclear receptors linked to adaptive thermogenesis. Cell 92:829–839

St Pierre J, Lin J, Krauss S et al 2003 Bioenergetic analysis of peroxisome proliferator-activated receptor γ coactivators 1α and 1β (PGC-1α and PGC-1β) in muscle cells. J Biol Chem 278:26597–26603

St Pierre J, Drori D, Uldry M et al 2006 Suppression of reactive oxygen species and neurodegeneration by the PGC-1 transcriptional coactivators. Cell 127:397–402

Sandri M, Lin J, Handschin C et al 2006 PGC-1alpha protects skeletal muscle from atrophy by suppressing FoxO3 action and atrophy-specific gene transcription. Proc Natl Acad Sci USA 103:16260–16265

Schreiber SN, Emter R, Hock MB et al 2004 The estrogen-related receptor alpha (ERRalpha) functions in PPARgamma coactivator 1alpha (PGC-1alpha)-induced mitochondrial biogenesis. Proc Natl Acad Sci USA 101:6472–6647

Tiraby C, Tavernier G, Lefort C et al 2003 Acquirement of brown fat cell features by human white adipocytes. J Biol Chem 278:33370–33376

Wu Z, Puigserver P, Andersson U et al 1999 Mechanisms controlling mitochondrial biogenesis and respiration through the thermogenic coactivator PGC-1. Cell 98:115–124

DISCUSSION

Nicholls: You talk about protection against MPTP and kainic acid. In the sorts of experiments that we do *in vitro* there is rapid damage occurring. So is PGC1α being induced fast enough? Are proteins being synthesized fast enough?

Spiegelman: Probably not for the first wave. The system has tone to begin with, but in terms of the most acute responses to an insult, I don't think it will be protected. This will take minutes to hours, not seconds to minutes. The initial tone matters. In our PGC1α manipulations we set the tone by having knockouts or by increased PGC1α but by no means is this the first means of defence. Martin Brand would probably argue that ROS activating uncoupling proteins is a second to minutes attribute, and therefore is positioned to respond a lot more quickly. This is clearly going to be a chronic response and one that sets tone to the system from the beginning.

Jacobs: You have struck gold with PGC1α, but where do PGC1β and PRC (PGC1-related coactivator) fit in? Are they acting in parallel, or antagonistically in some way?

Spiegelman: PGC1β is just as good as PGC1α in mitochondrial biogenesis. It tends not to be inducible, and it tends not to have some of the tissue-specific bells and whistles. PGC1α doesn't just turn on mitochondrial biogenesis, but turns on the whole fasting pathway in the liver including gluconeogenesis. PGC1α turns on mitochondrial biogenesis in brown fat but it also turns on uncoupling protein 1 (UCP1). With PGC1β you just get mitochondrial biogenesis. It looks like a constitutive basal function. From a regulatory point of view there are some things that induce β, but nothing in comparison with α. In some contexts, α induces β. In the induction of the anti-ROS programme, if you ablate α, β isn't induced. α turns on β and both of them probably co-operate going after gene sets.

Jacobs: This means that α requires β for its full activity. The prediction from this is that the β knockout will show a similar phenotype, if α requires β for protection against oxidative stress.

Spiegelman: Not necessarily. The β knockout does not have the neurodegeneration that the α does. In a lot of pathways PGC1β looks like an accessory.

Jacobs: What about PRC?

Spiegelman: Almost nothing is known about it. We have made a β transgenic mouse. It is bizarre but interesting. It has 100% IIX fibres (Arany et al 2007).

Rizzuto: What happens in neurons in a PGC1α-overexpressing mouse?

Spiegelman: It hasn't been looked at. We have one now in the substantia nigra but we are still characterizing it.

Rizzuto: Do any of the mice have an extended lifespan or fewer pathological alterations?

Spiegelman: If you tell me the experiment to do, I'd be interested in doing it! The problem is that the knockout has so many phenotypes that I'm not sure how to interpret the data. They seem to live well, but we are comparing a mouse that is hyperactive and lean to a mouse which ordinarily in its life gets fatter and develops insulin resistance. If I knew which tissue to look in I'd be able to do the tissue-specific knockout and let it age.

Scorrano: I wanted to address the effects on the PGC1α transgenic and the knockout. In the transgenic animal you get a blockage of muscular atrophy. It is highly debated whether atrophy is the expression of apoptosis in the muscle. On the other hand if you knock it out there is an increase in apoptosis induced by MPTP in the substantia nigra. Would you conclude that PGC1α somehow regulates apoptosis by regulating the mitochondrial pathways of apoptosis?

Spiegelman: That's an interesting idea; it is not what our paper is centred on. We collaborated with Fred Goldberg on this who is interested in muscle atrophy. One thing I can say is that PGC1 expression blocks the so-called 'atrogenes'—the E3 ubiquitin ligases that are specific for skeletal muscle, Murf and Atrogen. For sure, PGC1α blocks the expression of these. I'm sure many other things are happening. We didn't look at apoptosis in this experiment; we looked at shrinkage of the fibres and the expressions of the atrogenes.

Scorrano: There is increased apoptosis in the knockout when you treat with MPTP.

Spiegelman: Yes, and in the striatum when we don't do anything. The defect in the striatum is a basal defect. We don't have to do anything to the animals and they have a striatum that looks like Swiss cheese. This specific loss of neurons may be apoptosis. Bob Bachoohas looked at electron micrographs and his preliminary thought is that this is neural and not glial based.

Turnbull: What happens to the muscle in the PGC1-overexpressing mouse, with regards to glucose tolerance?

Spiegelman: Not much either way. The muscle is converting almost exclusively to the utilization of fatty acids and lipids. On balance, if we step back and ask about glucose homeostasis, it is not much changed. This is a bit surprising. We get about three times the mass of mitochondria in the muscles, but then we are not asking the mice to do any work.

Orrenius: I have a question about the antioxidant response. We know that antioxidant enzymes protect cells from apoptosis and other stresses, and that the mitochondrial isozymes are often more important for such protection than those located in the cytosol. Since you get such impressive protection by PGC1α induction, would you know if there is a selectivity for induction of the mitochondrial forms of the antioxidants?

Spiegelman: No, it is broad based. The reviewers of our paper raised the point that this should be expected: we are turning on everything else in the mitochondria, but we are actually turning on a lot of things that are not in the mitochondria, as well. As far as which are more important, I have no opinion. But it turns on mitochondrial antioxidant systems and non-mitochondrial antioxidant systems. It turns them on as a package.

Orrenius: With regards to the hydrogen peroxide results, it seems that the specific form of oxidant present in the mitochondria is particularly important. Have you been looking at this?

Spiegelman: We mainly looked at GPX. We have also used paraquat and MPTP. I think there is generality: this isn't a specific H_2O_2 response.

Brand: Following on from this, because the PGC1α is switching up both mitochondrial expression and ROS defences, presumably it is true that the ROS defensive subset of the mitochondrial expression goes up much more than the rest of that set. Is that true?

Spiegelman: That's a great question. There is something else going on. It is smartening what it does, either because the transcription factors are being co-induced with the ROS challenge, or PGC1α is being modified. If you stimulate PGC1 expression with ROS, it tends to go after the ROS systems more than the standard mitochondrial genes. If you take PGC1 under basal conditions and put it into your favourite cell, you will see the anti-ROS systems go up, but the standard mitochondrial systems will go up a lot more. There is definitely an education process going on. There will soon be a flurry of papers about the education of PGC1α by various protein kinases and systems. It is definitely smart. There is a bigger effect on SOD2 than on cytochrome c when PGC1α is induced, so much so that we are now double pegging PGC1α cells and animals, giving them an oxidative challenge and then purifying the complexes. I know where you are coming from in your question: there is something more going on than I alluded to. It definitely knows which way to go. It is a soft-wired programme that is sensitive to modification in terms of the relative levels of the different genes that are being expressed.

Reynolds: There is another pathway that unfortunately has the same name as one of your transcription factors, NRF2. When you talk about it, it sounds like the same thing: there is one pathway that turns on a whole bunch of antioxidant genes. Is the other NRF2 downstream of PGC1α?

Spiegelman: We are studying it now. In the lab we refer to NRF2 as GABP to alleviate the confusion.

Jacobs: A slight change of direction. Have you looked at the range of human variation in basal expression and inducibility of PGC1α?

Spiegelman: We came from the world of obesity and diabetes. There is a polymorphism at amino acid 462 that was discovered by Olof Petersen, which has been replicated numerous times, associated with human diabetes. There is also an interesting polymorphism in PGC1β. I assume the association studies will be done soon.

Jacobs: On the basis of your mouse results you could expect that humans with a more inducible PGC1α would be the ones that are more robust, and are consequently better athletes.

Spiegelman: That could be the case. A drug that could elevate PGC1α might be subject to abuse by athletes.

Nicholls: I like the idea of exercise without having to exercise!

Martinou: In the striatum of PGC1 knockouts, do the neurons contain a normal number of mitochondria?

Spiegelman: We haven't looked at mitochondrial number, just gene expression. It depends on the subunit, but in the PGC1α knockout in every tissue there is a mild to moderate decrease in mitochondrial gene expression.

Beal: Do you know anything about the time course of the degeneration in the brain?

Spiegelman: It's from the earliest stage we have looked, and it doesn't progress with time. At least in the striatum it is bad to start with. Those are good questions. People who look at the mice say that they look more like Huntington's disease than the Huntington mice.

Beal: There are two papers saying that resveratrol activates PGC1α, increasing mitochondrial numbers and exercise endurance (Baur et al 2006, Lagouge et al 2006). Is the answer to use resveratrol?

Spiegelman: Even if you accept what they say in the paper, this dosing in humans may not be possible. It shows the potential of the system to be manipulated.

Shirihai: Going back to diabetes, in the β cell do you see an effect of GLP1, for example, which causes hypertrophy of the β cell and proliferation of the mitochondria? Would this work through PGC1α?

Spiegelman: I don't know. Everything that induces cAMP is usually going to induce PGC1. GRP1 has a pro-growth, pro-health effect on the β cell. I don't really know the story.

Shirihai: β cells are low in antioxidant scavenging mechanisms. It would be interesting to see whether expression of PGC1α would still induce a strong antioxidant response, especially under conditions such as glucotoxicity and lipotoxicity.

Larsson: I have a question concerning down-regulated PGC1α expression in insulin resistance. A while ago I surveyed the literature on patients with mitochondrial defects, and they don't seem to be insulin resistant but they certainly have insulin secretion defects (see Wredenberg et al 2006 and references therein). We also did glucose homeostasis studies in a mouse with a selective respiratory chain deficiency in skeletal muscle. These mice have an increased uptake of glucose after a glucose tolerance test. If muscle is isolated and incubated with radioactive glucose, the uptake is increased. The finding that PGC1α is down-regulated in diabetic human skeletal muscle suggests that the respiratory chain deficiency is a secondary phenomenon. Acquired mitochondrial damage causing respiratory chain dysfunction will result in PCG1α up-regulation and increased mitochondrial biogenesis.

Spiegelman: That's a good, sophisticated question. What the field calls the ox-phos theory of diabetes I call the PGC1α theory of diabetes. PGC1 regulates a lot more than just the ox-phos system. The correlation is there. I don't assume that if this is causal, it is related to the regulation of ox-phos. It might: I have nothing against this idea, but we also show that PGC1α regulates Glut4 and fibre-type switching in the genes related to fibre types. Unlike most other people, I don't assume that the business end of this is the ox-phos system. I think the ox-phos system is a convenient tag for them to find that the PGC1 coactivators are down, because they do a lot more than just ox-phos.

Larsson: This supports my view that down-regulation of ox-phos capacity is likely a secondary phenomenon not primarily involved in causing insulin resistance in diabetic skeletal muscle.

Spiegelman: Vamsi Mootha probably wouldn't have noticed a down-regulation of the PGC1s at all except that the ox-phos was a nice tag for looking at the system.

O'Rourke: I was curious about the role of Akt in regulating the system. A recent presentation (Izumiya 2006) reported the generation of an Akt switch mouse. When Akt was turned on there was a tremendous growth of type 2 muscle fibres. They could reverse obesity and insulin resistance. They saw the opposite of what you are seeing: a switch from type 1 to type 2 fibres.

Spiegelman: It was presented at a meeting (but not by us) that Akt phosphorylation of PGC1α destroys it.

Reynolds: I have an off-the-wall question: what is biogenesis? Are there 'stem' mitochondria?

Spiegelman: The best correlate is mitochondrial gene expression and respiration. When you are talking about either gain of function or loss of function, in some cases it is increased numbers, in some cases it is not. In some cases we see increased volume without increased number; on others we see decreased volume without decreases in number. So we tend to use volume rather than numbers. But it is an interesting semantic issue: you picked up on the fact that I flung the biogenesis term around a little loosely. I don't necessarily mean increased number.

Reynolds: My particular interest in this question is that we are thinking about neurons, which are odd shaped cells compared with most other cells in the body. A neuron can be a metre long, and what goes on in one part of the cell could be quite different from what goes on elsewhere. So whether a new mitochondrion can be made locally is an issue here.

Duchen: What happens to mitochondrial DNA copy number? Could this be another measure of biogenesis?

Spiegelman: When we have looked at it we have seen that it goes up. We are loathe to use this as a marker, though.

References

Arany Z, Lebrasseur N, Morris C et al 2007 The transcriptional coactivator PGC-1beta drives the formation of oxidative type IIX fibers in skeletal muscle. Cell Metab 5:35–46

Baur JA, Pearson KJ, Price NL 2006 Resveratrol improves health and survival of mice on a high-calorie diet Nature 444:337–442

Izumiya Y 2006 Akt1-Mediated Muscle Growth Will Regress Adiposity and Insulin Resistance in Obese Mice. Manuscript presented in the Louis B. Katz Award competition at the American Heart Association Annual Meeting

Lagouge M, Argmann C, Gerhart-Hines Z 2006 Resveratrol Improves Mitochondrial Function and Protects against Metabolic Disease by Activating SIRT1 and PGC-1α Cell 127: 1109–1122

Wredenberg A, Freyer C, Sandstrom ME et al 2006 Respiratory chain dysfunction in skeletal muscle does not cause insulin resistance. Biochem Biophys Res Commun 350:202–207

Novel uncoupling proteins

Charles Affourtit, Paul G. Crichton, Nadeene Parker and Martin D. Brand[1]

MRC Dunn Human Nutrition Unit, Hills Road, Cambridge CB2 2XY, UK

Abstract. Mitochondria are incompletely coupled because of proton leaks that short-circuit oxidative phosphorylation. Basal proton leak is unregulated and is associated with the presence (but not catalytic activity) of the adenine nucleotide translocase. Inducible proton leak is regulated and is catalysed by the adenine nucleotide translocase and specific uncoupling proteins (UCPs). UCP1 catalyses proton conductance in mammalian brown adipose tissue. It is activated by fatty acids, which overcome nucleotide inhibition. UCP2, UCP3 and UCPs from birds, fish and plants also catalyse proton conductance that is inhibited by nucleotides. However, they require activation by superoxide or other reactive oxygen species (ROS). The mechanism of proton transport by the UCPs is unresolved. UCPs may also transport fatty acids or fatty acyl peroxides. Several physiological functions of UCPs are postulated. (1) UCP1 is specialised for thermogenesis; UCP3 and avian UCPs possibly share this function. (2) UCPs may attenuate ROS production and protect against oxidative damage, degenerative diseases and ageing. (3) UCP3 may catalyse fatty acid transport. (4) UCP2 has a signalling role in pancreatic β cells, where it attenuates insulin secretion. Other roles remain to be discovered.

2007 Mitochondrial biology: new perspectives. Wiley, Chichester (Novartis Foundation Symposium 287) p 70–91

Mitochondrial electron transfer and ATP synthesis are coupled incompletely due to proton leakage across the mitochondrial inner membrane. Up to 20% of the basal metabolic rate may be used to drive the proton leak, which indicates its physiological relevance (Brand 2005). Leak activity in isolated mitochondria is predominantly constitutive and depends to a large extent on the amount (but not activity) of the adenine nucleotide translocase (Brand et al 2005). Importantly, some proton leak is inducible, which allows regulation of energy transduction efficiency (Brand 2005). Inducible proton conductance is accounted for by the adenine nucleotide translocase and specific uncoupling proteins (UCPs). The archetypal uncoupling protein, UCP1, has been characterized in relative detail; its more recently discovered orthologues UCP2 and UCP3 as well as the avian, fish and plant UCPs are

---

[1] This paper was presented at the symposium by Martin Brand, to whom correspondence should be addressed.

understood less well. Despite the enormous interest in the roles of the novel UCPs, their possible physiological functions (Brand & Esteves 2005) and the reactions they catalyse (Esteves & Brand 2005) remain uncertain and are subject to ongoing debate (Nicholls 2006). In this paper we aim to provide a brief overview of current ideas on the mechanism and function of UCPs.

Mechanism of proton conductance

UCP catalysis

UCP1 catalyses the net transfer of protons across the mitochondrial inner membrane, dissipating the protonmotive force as heat. This proton leak activity is inhibited by purine nucleoside di- and triphosphates and stimulated by fatty acids (Nicholls & Rial 1999). The primary structure of novel UCPs is very similar to that of UCP1, and despite their relatively low abundance, they too mediate, under appropriate conditions (see below), proton conductance that is increased by fatty acids and is sensitive to nucleotides (Echtay et al 2002). Given these similarities, the mechanism of proton transport by novel UCPs is thought to be comparable to that of UCP1 (Esteves & Brand 2005).

Despite intensive research, mainly on isolated protein reconstituted in liposomes, the transport mechanism of UCP1 remains unresolved. It has been proposed that UCP1 transports protons directly using the carboxyl group of fatty acids as a co-factor to allow stepwise transfer across the membrane (Klingenberg & Echtay 2001). An alternative model explains the proton conductance of UCPs by the well-documented ability of mitochondrial carriers to transport anions: protonated fatty acids flip across the membrane protein-independently and are transported back by UCPs as fatty acid anions (Skulachev 1991). More recent models suggest that proton transport is independent of fatty acids, which act only as regulators of conductance (Nicholls 2006, Shabalina et al 2004). The lack of proton conductance associated with the UCP1-dependent transport of alkylsulfonates (fatty acid analogues that are unable to flip) has been cited as the strongest evidence for the 'cycling' model, although several other lines of evidence do not support it (Klingenberg & Echtay 2001). Most prominent is the identification of a site-specific UCP1 mutant that retains the ability to transport protons yet is unable to transport anions, indicating that anion movement is not required for proton transport (Echtay et al 2000a).

The mechanism by which UCP activity is inhibited by nucleotides is also unclear. In mitochondria isolated from brown adipose tissue, physiological nucleotide concentrations are sufficient to retain UCP1 in a fully inhibited state, which can be overcome by fatty acids. The interaction between fatty acids and nucleotides appears functionally competitive (Shabalina et al 2004). Whether or not UCP1 catalyses a proton conductance in the complete absence of these effectors remains controversial as the unambiguous removal of fatty acids from isolated mitochondria is difficult to verify (Klingenberg & Echtay 2001). It is worth noting, however, that in proteoliposomes without nucleotides, the proton conductance activity of UCP1, UCP2, and UCP3 is fully dependent on the presence of fatty acids (Echtay et al 2000b, 2001, Jaburek & Garlid 2003), and is non-competitively inhibited by added nucleotides (Zackova et al 2003).

UCP activation

Novel UCPs do not contribute to basal mitochondrial proton conductance in the absence of specific activators (Esteves & Brand 2005). However, superoxide (generated artificially or formed endogenously) induces a nucleotide-sensitive proton conductance in mitochondria from a variety of tissues. Such induction correlates with the presence of novel UCPs in mammalian (Cadenas et al 2002, Echtay et al 2002, Krauss et al 2003), bird (Talbot et al 2004) and plant systems (Considine et al 2003), and is not observed in kidney mitochondria isolated from *Ucp2* knockout mice (Krauss et al 2003) or skeletal muscle mitochondria from *Ucp3* knockout mice (Cadenas et al 2002). Superoxide also stimulates UCP1 activity, both in mitochondria from brown adipose tissue and recombinant yeast expressing the protein (Echtay et al 2002).

How does superoxide activate UCPs? The main mechanistic model predicts that activation is indirect, and involves superoxide-induced lipid peroxidation (Brand & Esteves 2005). A major breakdown product of this process is hydroxynonenal (HNE), which has been shown to induce nucleotide-sensitive proton conductance mediated by mammalian (Echtay et al 2003) and plant (Smith et al 2004) UCPs. When mitochondria are incubated with mitoPBN, a mitochondrially targeted scavenger of carbon-centred radicals, activation of UCP by both superoxide and chemical generators of carbon-centred radicals is prevented. Importantly, however, mitoPBN does not interfere with UCP activation by HNE (Murphy et al 2003). These results suggest that superoxide activates UCPs through generation of carbon-centred radicals leading to formation of HNE, which could be the species that interacts with the protein (Brand & Esteves 2005, Murphy et al 2003).

How HNE interacts with UCPs is unclear, although its general ability to form protein adducts with various amino acids (a common sign of oxidative damage) may account for the observed activation (Brand et al 2004). HNE and fatty acids are likely to induce UCP activity by separate mechanisms, as evidenced by their synergistic effect on UCP1 activity in recombinant yeast mitochondria (Esteves et al 2006). As with catalysis, however, UCP activation by HNE is controversial (Shabalina et al 2006). Equally, different mechanisms of superoxide-induced activation of UCP have been proposed. One theory predicts that lipid peroxidation caused by superoxide leads to formation of hydroperoxy fatty acids, which have been shown to support UCP2-mediated uncoupling in proteoliposomes (Jaburek et al 2004).

Physiological function

The role of novel UCPs in thermogenesis

During cold exposure of rodents, UCP1 levels can reach up to 6% of the total mitochondrial inner membrane protein content of brown adipose tissue, an organ highly involved in heat production during non-shivering thermogenesis (Stuart et al 2001). Moreover, *Ucp1* knockout mice are hindered severely in their ability to cope with cold exposure (Cannon & Nedergaard 2004), showing that UCP1 plays an important role in adaptive thermogenesis. Based on considerable sequence similarity, it was initially supposed that UCP2 and UCP3 would have a thermogenic function too. Although this remains an attractive idea, there is mounting evidence refuting it.

Both UCP2 and UCP3 are up-regulated in skeletal muscle of fasted rats, but thermogenesis is decreased rather than increased (Boss et al 1997). This phenotype is consistent with the observation that isolated skeletal muscle mitochondria lacking UCP3 exhibit identical basal proton conductance to those expressing it at physiological levels (Cadenas et al 1999). The most compelling evidence against a thermogenic role for the novel UCPs comes from the lack of a strong metabolic phenotype in *Ucp2* and *Ucp3* knockout mice, which are neither obese nor compromised in adaptive thermogenesis (Harper & Himms-Hagen 2001). Nonetheless, changes in novel UCP content correlate with thermogenesis under certain conditions, and in some species that do not have brown adipose tissue. For instance, *Ucp3* knockout mice are protected against hyperthermia caused by administration of ecstasy (MDMA), suggesting that UCP3 is central to the thermogenic response (Mills et al 2003). Furthermore, cold exposure causes up-regulation of avian UCP in skeletal muscle of birds (Mozo et al 2005, Talbot et al 2004) and UCP1 in the liver of fish (Jastroch et al 2005).

That novel UCPs do not contribute generally to basal or cold-stressed metabolism in rodent models suggests that adaptive thermogenesis is not their evolutionary function. It can be argued that this is because these proteins are either thermogenically inactive or not sufficiently abundant for their effects to be observed in whole animals. However, under certain conditions (e.g. MDMA treatment) activation of novel UCPs can contribute significantly to thermogenesis. UCPs therefore remain an attractive target for the treatment of metabolic imbalance.

The role of novel UCPs in the attenuation of reactive oxygen species production

Reactive oxygen species (ROS) are a major cause of oxidative stress and consequent cellular damage and have been associated with many degenerative diseases (Brand et al 2004). The discovery that the mitochondrial proton conductance mediated by novel UCPs is activated potently by superoxide (Echtay et al 2002) has raised the exciting possibility that this group of proteins may protect against such diseases.

ROS are inevitable products of mitochondrial electron transport, and their production is exquisitely sensitive to the protonmotive force. Superoxide-stimulated UCP activity uncouples oxidative phosphorylation mildly, which results in a drop in protonmotive force that would diminish ROS formation, but would be sufficiently small to allow continued ATP synthesis. Activation of UCPs by superoxide would thus provide an elegant negative feedback loop (Fig. 1) to dissipate harmful ROS production and attenuate oxidative stress (Brand et al 2004).

In addition to strong direct evidence obtained *in vitro* (Brand & Esteves 2005, Esteves & Brand 2005), several results from studies involving mouse knockout models provide support for a protective function of UCPs. For example, *Ucp2*-ablated mice exhibit an approximately 32% increase in ROS in macrophages (Arsenijevic et al 2000) and have accelerated atherosclerotic plaque development (Blanc et al 2003). Therefore, UCP2 activity appears to attenuate ROS in

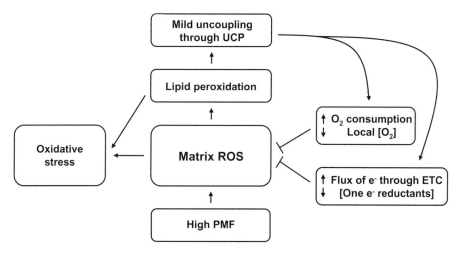

FIG. 1. Protective role of UCPs. Mitochondrial ROS are produced by the electron transfer chain (ETC) when protonmotive force (PMF) is high. Mild uncoupling lowers ROS and consequently oxidative stress. Uncoupling proteins (UCPs) provide a negative feedback to attenuate high ROS levels. ROS promote lipid peroxidation leading to formation of HNE, which activates UCPs.

macrophages, conferring protection against progression of atherosclerosis. A negative (secondary) consequence of relatively low ROS levels, however, could be decreased toxoplasmacidal and bactericidal activity in macrophages: *Ucp2* knockout mice are resistant to normally lethal *Toxoplasma gondii* infection, and this effect disappears upon addition of a ROS scavenger (Arsenijevic et al 2000). Furthermore, skeletal muscle mitochondria isolated from *Ucp3* knockout mice have significantly higher ROS levels and oxidative damage than wild-type controls (Brand et al 2002, Vidal-Puig et al 2000), suggesting that mild uncoupling may be protective.

Hence, UCPs appear to act as part of the cell's antioxidant defence. Whilst this appears to be useful for the long-term protection against ROS, it seems that short-term requirements for elevated ROS by the immune system are compromised.

Export of fatty acids by UCP3

When substrate supply exceeds β oxidation capacity, fatty acids might accumulate in the mitochondrial matrix. If they enter and build up as fatty acyl CoA esters, it is claimed that exhaustion of free CoA might limit further oxidation, whereas accumulation as free acids (if entry occurs via a flip-flop mechanism) might be toxic. In two closely related functional models, UCP3 has been proposed to export fatty acid anions from the mitochondrial matrix, either to allow high fatty acid oxidation rates (Himms-Hagen & Harper 2001) or to protect against lipotoxicity (Schrauwen et al 2006). A role in fatty acid metabolism is supported by correlations between expression of UCP3 and genes involved in fatty acid oxidation. Furthermore, UCP3 expression is up-regulated by long-chain fatty acids and tends to correlate with their oxidation. Such observations, however, are equally well accommodated by a model that predicts that UCP3 attenuates ROS production by mild uncoupling of oxidative phosphorylation (Brand & Esteves 2005). More direct evidence is required to distinguish between the models.

Regulation of insulin secretion by UCP2

The main mechanism of glucose-stimulated insulin secretion (GSIS) in pancreatic β cells (Rutter 2001) relies on efficient mitochondrial energy transduction. Because of such efficiency, the cytoplasmic ATP/ADP ratio in β cells rises when their oxidative glucose catabolism increases in response to elevated blood glucose levels. This subsequently causes closure of K_{ATP} channels, membrane depolarization, opening of voltage-sensitive calcium channels, calcium influx, and secretion of insulin (Fig. 2). Mitochondria isolated from cultured β cells exhibit proton conductance that is stimulated considerably by superoxide in a GDP-sensitive manner (Echtay et al 2002). Given that UCP2 mRNA is expressed in pancreatic islets (Zhou et al 1997), this suggests that β cells contain functional UCP2 protein that can be activated by ROS.

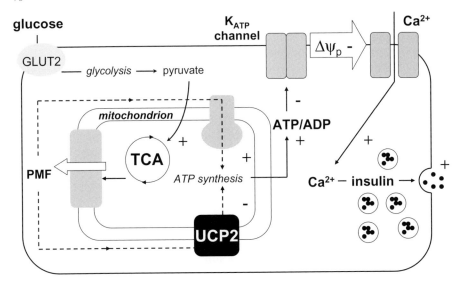

FIG. 2. Role of UCP2 in glucose-stimulated insulin secretion. Glucose enters the β cell via a transporter (GLUT2). Glucose catabolism establishes a mitochondrial protonmotive force (PMF) that drives ATP synthesis, increasing cytosolic ATP/ADP. This causes closure of K_{ATP} channels, depolarisation of the plasma membrane potential ($\Delta\psi_p$), opening of voltage-dependent calcium channels, calcium influx and insulin secretion. Coupling between glucose oxidation and ATP synthesis is decreased by UCP2 activity, which dissipates the protonmotive force.

Given its dissipative effect on the mitochondrial protonmotive force (Fig. 2), it is conceivable that UCP2 activity in β cells attenuates the response of ATP/ADP to glucose, and modulates insulin secretion. Persuasive evidence for this signalling role comes from loss of function studies. Pancreatic islets from *Ucp2*-ablated mice have higher ATP levels than wild-type controls and exhibit increased GSIS, showing that UCP2 regulates insulin secretion negatively (Zhang et al 2001). In wild-type islets, fatty acids diminish glucose-induced increases in mitochondrial membrane potential, ROS production, ATP/ADP ratio, cytosolic calcium and insulin secretion. This attenuating effect of fatty acids is not seen in UCP2-deficient islets, implicating UCP2 as a mediator (Joseph et al 2004). Furthermore, removal of endogenously produced superoxide with a superoxide dismutase mimetic improves GSIS. Hyperglycaemia impairs GSIS through a mechanism that involves superoxide. Both these superoxide effects are apparent exclusively in islets from wild-type animals, again corroborating a signal-mediating role for UCP2 (Krauss et al 2003). Recent additional support for a regulatory role of UCP2 in GSIS comes from studies with genipin, a cell-permeant molecule that inhibits UCP2-mediated proton leak in kidney mitochondria (Zhang et al 2006). In islets, genipin increases the mitochondrial membrane potential and cytosolic ATP levels, closes K_{ATP} channels, and stimulates insulin secretion; these effects are UCP2-dependent. Moreover, genipin acutely reverses glucose- and obesity-induced β cell dysfunction.

Based on these observations, UCP2 has been implicated in β cell dysfunction and the development of type 2 diabetes (Lowell & Shulman 2005). It may be that UCP2-related pathology is secondary to a defence mechanism that restricts ROS production and islet damage that would result from a high-fat and high-glucose diet (Brand & Esteves 2005). Alternatively, acute attenuation of GSIS by UCP2 may be important physiologically, perhaps to regulate responses to variable nutrient supply (Brand et al 2004).

Glucose-dependent fluctuations in ATP/ADP are observed in β cells but not in other cell types such as muscle, where regulatory feedback loops ensure this poise is relatively unresponsive to substrate supply. The difference in flexibility arises partly from mitochondrial particularities (Affourtit & Brand 2006). Proton leak is higher in β cell mitochondria than muscle mitochondria. Consequently, the control exerted by leak over ATP/ADP is relatively strong in β cell mitochondria. Furthermore, leak has relatively high control over membrane potential and respiratory activity. Importantly, leak differences are observed at the cellular level too. RNAi experiments demonstrate that UCP2 contributes significantly to the relatively high leak activity in intact β cells (Affourtit & Brand, unpublished). The central role of proton leak in β cell energetics supports the potential physiological importance of UCP2.

Concluding remarks

In the last 10 years more than 1000 papers have been published with 'uncoupling protein' or 'UCP' in the title. Although we understand neither their mechanism nor their full functional significance, study of the novel uncoupling proteins is an exciting and fast-moving area that promises to produce many more surprises and insights in the next 10 years.

References

Affourtit C, Brand MD 2006 Stronger control of ATP/ADP by proton leak in pancreatic beta-cells than skeletal muscle mitochondria. Biochem J 393:151–159
Arsenijevic D, Onuma H, Pecqueur C et al 2000 Disruption of the uncoupling protein-2 gene in mice reveals a role in immunity and reactive oxygen species production. Nat Genet 26:435–439
Blanc J, Alves-Guerra MC, Esposito B et al 2003 Protective role of uncoupling protein 2 in atherosclerosis. Circulation 107:388–390
Boss O, Samec S, Dulloo A, Seydoux J, Muzzin P, Giacobino JP 1997 Tissue-dependent upregulation of rat uncoupling protein-2 expression in response to fasting or cold. FEBS Lett 412:111–114

Brand MD 2005 The efficiency and plasticity of mitochondrial energy transduction. Biochem Soc Trans 33:897–904

Brand MD, Esteves TC 2005 Physiological functions of the mitochondrial uncoupling proteins UCP2 and UCP3. Cell Metab 2:85–93

Brand MD, Pamplona R, Portero-Otin M et al 2002 Oxidative damage and phospholipid fatty acyl composition in skeletal muscle mitochondria from mice underexpressing or overexpressing uncoupling protein 3. Biochem J 368:597–603

Brand MD, Affourtit C, Esteves TC et al 2004 Mitochondrial superoxide: production, biological effects, and activation of uncoupling proteins. Free Radic Biol Med 37:755–767

Brand MD, Pakay JL, Ocloo A et al 2005 The basal proton conductance of mitochondria depends on adenine nucleotide translocase content. Biochem J 392:353–362

Cadenas S, Buckingham JA, Samec S et al 1999 UCP2 and UCP3 rise in starved rat skeletal muscle but mitochondrial proton conductance is unchanged. FEBS Lett 462:257–260

Cadenas S, Echtay KS, Harper JA et al 2002 The basal proton conductance of skeletal muscle mitochondria from transgenic mice overexpressing or lacking uncoupling protein-3. J Biol Chem 277:2773–2778

Cannon B, Nedergaard J 2004 Brown adipose tissue: function and physiological significance. Physiol Rev 84:277–359

Considine MJ, Goodman M, Echtay KS et al 2003 Superoxide stimulates a proton leak in potato mitochondria that is related to the activity of uncoupling protein. J Biol Chem 278:22298–22302

Echtay KS, Winkler E, Bienengraeber M, Klingenberg M 2000a Site-directed mutagenesis identifies residues in uncoupling protein (UCP1) involved in three different functions. Biochemistry 39:3311–3317

Echtay KS, Winkler E, Klingenberg M 2000b Coenzyme Q is an obligatory cofactor for uncoupling protein function. Nature 408:609–613

Echtay KS, Winkler E, Frischmuth K, Klingenberg M 2001 Uncoupling proteins 2 and 3 are highly active H^+ transporters and highly nucleotide sensitive when activated by coenzyme Q (ubiquinone). Proc Natl Acad Sci USA 98:1416–1421

Echtay KS, Roussel D, St-Pierre J et al 2002 Superoxide activates mitochondrial uncoupling proteins. Nature 415:96–99

Echtay KS, Esteves TC, Pakay JL et al 2003 A signalling role for 4-hydroxy-2-nonenal in regulation of mitochondrial uncoupling. EMBO J 22:4103–4110

Esteves TC, Brand MD 2005 The reactions catalysed by the mitochondrial uncoupling proteins UCP2 and UCP3. Biochim Biophys Acta 1709:35–44

Esteves TC, Parker N, Brand MD 2006 Synergy of fatty acid and reactive alkenal activation of proton conductance through uncoupling protein 1 in mitochondria. Biochem J 395:619–628

Harper ME, Himms-Hagen J 2001 Mitochondrial efficiency: lessons learned from transgenic mice. Biochim Biophys Acta 1504:159–172

Himms-Hagen J, Harper ME 2001 Physiological role of UCP3 may be export of fatty acids from mitochondria when fatty acid oxidation predominates: an hypothesis. Exp Biol Med 226:78–84

Jaburek M, Garlid KD 2003 Reconstitution of recombinant uncoupling proteins: UCP1, -2, and -3 have similar affinities for ATP and are unaffected by coenzyme Q10. J Biol Chem 278:25825–25831

Jaburek M, Miyamoto S, Di Mascio P, Garlid KD, Jezek P 2004 Hydroperoxy fatty acid cycling mediated by mitochondrial uncoupling protein UCP2. J Biol Chem 279:53097–53102

Jastroch M, Wuertz S, Kloas W, Klingenspor M 2005 Uncoupling protein 1 in fish uncovers an ancient evolutionary history of mammalian nonshivering thermogenesis. Physiol Genomics 22:150–156

Joseph JW, Koshkin V, Saleh MC et al 2004 Free fatty acid-induced beta-cell defects are dependent on uncoupling protein 2 expression. J Biol Chem 279:51049–51056

Klingenberg M, Echtay KS 2001 Uncoupling proteins: the issues from a biochemist point of view. Biochim Biophys Acta 1504:128–143

Krauss S, Zhang CY, Scorrano L et al 2003 Superoxide-mediated activation of uncoupling protein 2 causes pancreatic beta cell dysfunction. J Clin Invest 112:1831–1842

Lowell BB, Shulman GI 2005 Mitochondrial dysfunction and type 2 diabetes. Science 307:384–387

Mills EM, Banks ML, Sprague JE, Finkel T 2003 Pharmacology: uncoupling the agony from ecstasy. Nature 426:403–404

Mozo J, Emre Y, Bouillaud F, Ricquier D, Criscuolo F 2005 Thermoregulation: what role for UCPs in mammals and birds? Biosci Rep 25:227–249

Murphy MP, Echtay KS, Blaikie FH et al 2003 Superoxide activates uncoupling proteins by generating carbon-centered radicals and initiating lipid peroxidation: studies using a mitochondria-targeted spin trap derived from alpha-phenyl-N-tert-butylnitrone. J Biol Chem 278:48534–48545

Nicholls DG 2006 The physiological regulation of uncoupling proteins. Biochim Biophys Acta 1757:459–466

Nicholls DG, Rial E 1999 A history of the first uncoupling protein, UCP1. J Bioenerg Biomembr 31:399–406

Rutter GA 2001 Nutrient-secretion coupling in the pancreatic islet beta-cell: recent advances. Mol Aspects Med 22:247–284

Schrauwen P, Hoeks J, Hesselink MK 2006 Putative function and physiological relevance of the mitochondrial uncoupling protein-3: involvement in fatty acid metabolism? Prog Lipid Res 45:17–41

Shabalina IG, Jacobsson A, Cannon B, Nedergaard J 2004 Native UCP1 displays simple competitive kinetics between the regulators purine nucleotides and fatty acids. J Biol Chem 279:38236–38248

Shabalina IG, Petrovic N, Kramarova TV, Hoeks J, Cannon B, Nedergaard J 2006 UCP1 and defense against oxidative stress. 4-Hydroxy-2-nonenal effects on brown fat mitochondria are uncoupling protein 1-independent. J Biol Chem 281:13882–13893

Skulachev VP 1991 Fatty acid circuit as a physiological mechanism of uncoupling of oxidative phosphorylation. FEBS Lett 294:158–162

Smith AM, Ratcliffe RG, Sweetlove LJ 2004 Activation and function of mitochondrial uncoupling protein in plants. J Biol Chem 279:51944–51952

Stuart JA, Harper JA, Brindle KM, Jekabsons MB, Brand MD 2001 A mitochondrial uncoupling artifact can be caused by expression of uncoupling protein 1 in yeast. Biochem J 356:779–789

Talbot DA, Duchamp C, Rey B et al 2004 Uncoupling protein and ATP/ADP carrier increase mitochondrial proton conductance after cold adaptation of king penguins. J Physiol 558:123–135

Vidal-Puig AJ, Grujic D, Zhang CY et al 2000 Energy metabolism in uncoupling protein 3 gene knockout mice. J Biol Chem 275:16258–16266

Zackova M, Skobisova E, Urbankova E, Jezek P 2003 Activating omega-6 polyunsaturated fatty acids and inhibitory purine nucleotides are high affinity ligands for novel mitochondrial uncoupling proteins UCP2 and UCP3. J Biol Chem 278:20761–20769

Zhang CY, Baffy G, Perret P et al 2001 Uncoupling protein-2 negatively regulates insulin secretion and is a major link between obesity, beta cell dysfunction, and type 2 diabetes. Cell 105:745–755

Zhang CY, Parton LE, Ye CP et al 2006 Genipin inhibits UCP2-mediated proton leak and acutely reverses obesity- and high glucose-induced beta cell dysfunction in isolated pancreatic islets. Cell Metab 3:417–427

Zhou YT, Shimabukuro M, Koyama K et al 1997 Induction by leptin of uncoupling protein-2 and enzymes of fatty acid oxidation. Proc Natl Acad Sci USA 94:6386–6390

DISCUSSION

Spiegelman: I'd like to start with a discussion of some of the basics. You refer to uncoupling as being able to counter reverse electron transport through proton leak. Why do you refer to it as 'reverse electron transport' as opposed to being a steady state system, such that if you simply accelerate the flow of electrons you could reduce the $t_{1/2}$ of all the reduced components? In a steady state system why would you need reversed electron transport?

Brand: We have done a reasonable amount of work looking as ROS production by isolated mitochondria. We find the greatest ROS production we can get, perhaps 10 times higher than any other site, is when we drive electrons backwards into complex I from the quinone pool. This is what I was referring to by 'reverse electron transport'. We could discuss whether this ever happens in cells, or whether something that approximates it happens in cells which might be a steady-state backed-up system at reasonably high protonmotive force, where there isn't much net flow, but electrons are just dribbling back into the mechanism. I know some people think this never happens in cells, so are we just looking at the wrong thing? Then we could look at the subsidiary sites which are much lower in rate, and the potential sensitivity of these. These are potential-sensitive, but less so. The argument would work for those sites, but less strongly. This is one of my many back-up positions: if it is not reverse electron transport, it would still work, but not so effectively for forward electron transport or at other places in the chain.

Spiegelman: So uncoupling works, regardless of whether you believe in back-flow.

Brand: Yes, but it works better if there is reverse electron transport.

Nicholls: My take on this is as follows. There are three disparate complexes which are pumping protons and translocating electrons, and amazingly they are more or less matched in terms of how powerful they are, but it seems that complex I is thermodynamically the weakest. Bypassing complex I with α-glycerophosphate or succinate feeding straight into the quinone pool, you get perhaps 5 or 10 mV more of protonmotive force. This throws you over the threshold where you are producing a lot of ROS, whereas if you have to go through complex I, as happens in real cells, this hyperpolarized condition is not generated. Thus this high voltage dependency is not seen.

Spiegelman: On the other hand, fats don't use complex I as much.

Nicholls: Fats are half and half. The first oxidation is the flavoprotein one, and the second is NAD-linked. It is only with the artificial system involving succinate that there is no NAD-linked oxidation required. As soon as you go into a cell, succinate is coming from α-ketoglutarate via a NAD-linked step. In a real cell you are always restricted by the involvement of complex I.

Brand: We find that *in vitro*, fatty acid oxidation is a good way to get high ROS production. The neuronal cells that David Nicholls is talking about don't have fatty acid oxidation, at least for use as a fuel. This may be something that happens more in muscle, a tissue that really uses fatty acid as a fuel.

Spiegelman: Do you think this can help to explain why nerves love glucose and don't like fatty acids as a fuel source?

Brand: Yes.

Giulivi: In your model, it is proposed that protonated superoxide anion activates the lipid peroxidation pathway, and then you get activation of UCP. The pK_a of the superoxide anion is 4.8, so it will be difficult to get it protonated at 7 or 7.4 to initiate anything.

Brand: What is the pH in the membrane?

Giulivi: There has to be a pH gradient: it has to be more acidic outside.

Brand: This is in the hydrophobic core of the membrane where the favoured form of superoxide is the protonated form.

Giulivi: But then the pK_a of that species is 4.8. Say the pH is 7 outside and 8 inside, how do you initiate the whole thing with such a small gradient? Why do you have to go through lipid peroxidation initiated by these protonated superoxide anions? Why isn't there initiation just by modifying UCP by ROS oxidative damage? I don't understand why all these steps in between are needed.

Brand: In our model the superoxide doesn't see the aqueous phases, so the pH in these aqueous phases doesn't matter. The solubility of the superoxide anion in the membrane is very low, whereas the solubility of the protonated form is higher. This is the form that will be in the membrane, regardless of the pK_a in the aqueous phases and the pH on either side of the membrane. I presented what we find and the best model we can come up with. If I were to explain it teleologically, this mechanism has been selected because it works. If we make ROS, it damages the membrane and all we have to do is pick up an end product of damage to signal that something horrible is going on upstream. It may be inelegant, but it is biology.

Giulivi: If you put purified UCP into liposomes that have been peroxidized versus liposomes that haven't, do you see activation of UCP?

Brand: You can argue either way, depending on how you read the literature. There are experiments that say that UCP placed in a phospholipid bilayer works fine and no ROS addition is needed. Other experiments say it simply doesn't work unless you add ubiquinone. Our interpretation is that ubiquinone is somehow generating the radicals, and it is only in those systems that are not very oxidised that we see the requirement for the ubiquinone. This is obviously hand-waving, so we are trying to set this up and repeat it in the lab at the moment. It is easier to say than it is to do.

Adam-Vizi: In your model, superoxide initiated lipid peroxidation which in turn activates UCP. If one thinks in terms of physiology, it is difficult to imagine how the whole process will be terminated. Something should terminate it if it is a physiologically relevant mechanism, but once lipid peroxidation is involved, which is a non-enzymatic autocatalytic reaction, the process would be kept going.

Brand: It's an important question that we are trying to answer. There are many possibilities. One would be that this is actually a memory system. If you have a burst of oxidative damage, you switch on UCP and it stays switched on even after the oxidative burst goes away. This could precondition against a second burst of oxidative damage. However, another possibility is that the protein is simply recognized as changed and is pulled out of the membrane and degraded. If the initial oxidative insult has gone away, the modified protein would be cleared. We are trying to find out the basic characteristics of deactivation at the moment.

Adam-Vizi: I have a comment on the inhibition of ROS production by uncoupling. As David Nicholls mentioned, if succinate or α-glycerophosphate is used to support respiration, ROS production is membrane-potential dependent. With glutamate/malate it is much less so. If we go one step further to the complex physiology measuring ROS production in *in situ* mitochondria, using synaptosomes, then there is no membrane potential dependence of ROS production at all. The basal membrane potential in *in situ* mitochondria is already below the range where ROS production would be dependent on the membrane potential.

Brand: My defence is tissue specificity. I would argue that this probably doesn't happen much in neuronal tissues, because they don't oxidize fatty acids. Synaptosomes don't normally run this high ROS production, so the defence isn't there or if the protein is there it is for some of these other signalling purposes. I don't deny this is a weakness of the model.

Larsson: Wouldn't it be a good idea to try to purify uncoupling proteins from situations where you think there would be a lot of activation by ROS, and in states where there is no activation. This strategy would allow you to identify oxidative modifications of the uncoupling proteins.

Brand: We have spent a lot of time trying to sort this out, without much success. Our model is that there is a covalent activation. We have been doing mass spectrometry, pulling out the protein from activated mitochondria and non-activated controls to see whether there is some covalent modification.

Beal: People have been using UCP2 overexpressing mice and knockouts, and they are protected against neuronal insults such as ischaemia/reperfusion. They are more vulnerable if they lack UCP2. There is no question that UCP2 plays a role in the nervous system protecting against reactive oxygen species. There are reports of UCP4 and 5 in the nervous system. Are these invalid?

Brand: With UCPs you have to be wary of positive artefacts. If a UCP is misexpressed, it uncouples. It is clear that a lot of the early literature with UCP2 and

3 was confused by secondary uncoupling artefacts. The phenotype expected was obtained, but without the regulation, probably due to misfolding. I suspect that a lot of the overexpression *in vivo* is an uncoupling artefact that is giving the desired response, but it is not necessarily related to the native function of the protein. With UCP4 and 5 there is only one paper (Fridell et al 2004) that suggests that anyone can express UCP5 transgenically and see regulatable function. My position on UCP4 and 5 is that the jury is out. There is no other good evidence that they can be regulated. The sequences say that they are just other mitochondrial anion carriers.

Beal: We are very interested in coenzyme Q10. It has been reported to regulate uncoupling proteins. You suggested that it might be an oxidative effect, but almost all other data suggest that it is anti-oxidant, particularly in lipid membranes.

Brand: It can be both pro- and antioxidant depending on the redox state and which other oxidants are present. We would have to postulate that under the conditions of that experiment it was acting as a prooxidant, generating ROS.

Lemasters: In your flux voltage curves after activation, on the one hand some of the curves show a shift in the inflexion point, suggesting that the activated activity is still voltage-dependent. On the other hand, in other curves it all shifted downwards, suggesting that the activated proton channel was voltage-independent. I'm interested in which of the UCP activities are voltage-dependent and which are not. If they are voltage-dependent and you are below that dependence, then it isn't going to have any effect on ROS formation.

Brand: I have a two-part answer. One is that our theoretical understanding is that there isn't an inflexion point; this is an exponential dependence on voltage. It will always be potential sensitive with the same characteristics. The other is that it turns out that there are some quite complicated time-dependent effects in these experiments. If we titrate we are not just titrating potential but also in time. We are getting secondary activations which are distorting the shapes of the curves. Now that we understand this, we can control for time and get much cleaner data in terms of the shapes of the curves and the way that they are clearly nested exponentials. From the published work we are now looking at, we can see that the shapes are going a bit awry. We don't believe the shapes we have published quite as much as we did, although the effects we think are still valid. So all of the UCP activity is voltage-dependent, but it is more apparent at higher potentials.

Nicholls: With UCP1 in the presence of nucleotide, we found that as one titrates in fatty acid, the break point decreases (Nicholls 1977). So instead of the conductance increasing rapidly at say 200 mV, which gives you respiratory control, it will increase to 180 mV. One rather nice thing is that the conductance curve seems to clamp the membrane potential and protonmotive force at a particular suboptimal level, so that you are not getting wild fluctuations of potential as the respiratory rate changes. You are still generating a physiologically relevant ADP:ATP ratio when you have lost respiratory control, which is important in the brown fat context. ATP is still being produced by the mitochondria.

O'Rourke: I have a general technical question about the interpretation of the flux force relationship being due to proton leak. Have you looked at the ion selectivity of the change in VO_2 as a function of membrane potential? You are interpreting this as being due to proton movement. But when you put the UCPs in bilayers they can be ion channels too. If other ions are flowing, this will affect the overall VO_2. What is the selectivity of this pathway?

Brand: Before the novel uncoupling proteins were thought of, we did extensive work on the basal proton conductance. We are confident that it is mostly proton flow. We haven't redone those experiments with the UCP-dependent uncoupling.

O'Rourke: My concern is that in the presence of ROS you may be activating other permeation pathways.

Nicholls: In order to increase respiration, it has to be ultimately linked to protons. It is the rate of proton cycling around the proton circuit that defines the rate of respiration. Incidentally, with UCP1 our original hypothesis was that we are not dealing with protons at all but hydroxyl ions. UCP also translocates chloride in much the same sort of way.

Brand: That has been raised with K^+ cycling. It could be that UCPs are transporting K^+, and we are adding nigericin, which is a K/H exchanger. The net result would be proton flow through a K^+ dependence. We have tested this by leaving out K^+ and it doesn't make any difference.

Reynolds: The carboxy atractylate binding in the UCP3 knockouts is interesting. Is the right interpretation of what you said that the carboxy atractylate is binding to UCP3, or is it telling us that the adenine nucleoside translocase is making a complex with UCP3 in the way that it functions? Can you tell the difference between these possibilities?

Brand: We don't know the answer. That observation needs to be demonstrated by some other technique. We have no idea whether it is direct.

Nicholls: Stuck up in my office is a poor resolution tube gel which was the original one where we identified UCP1. In this we were using 8-azido-ATP as a photoaffinity label. In the absence of carboxyatractylate we labelled with the azido ATP a 30 and 32 kDa band. We included carboxyatractylate which blocked the 30 kDa band, leaving us with the 32 kDa band. If atractylate is acting on both proteins, it has very different actions. It will compete for nucleotide binding to the translocator, but it doesn't compete for nucleotide binding to UCP1.

Brand: UCP1 may be different from the others, or it may be that it is all going through some interaction. There is so much more UCP1 that it overrides the interaction. We don't know the answer.

Halestrap: What about bongkrekic acid: does it do the same as carboxyatractyloside?

Brand: It does the same.

Nicholls: The criticism has been raised that most of the superoxide activation effects you are getting are with very large exogenously added levels of superoxide. How do you respond to that?

Brand: When we make superoxide in the external aqueous phase it will be difficult to get large amounts into the membrane, whereas in the physiological situation we would argue that it is being made in the membrane at the site where it will do the damage, so much smaller amounts are needed.

Nicholls: Is there anything special about the ANT or is it just that it is present at 100 times higher concentration than all the other transporters? In other words, is it simply because there is an awful lot of ANT that it seems to contribute to the activatable, nucleotide-inhibitable proton conductance, or is there some specific property to it?

Brand: Our hypothesis is that it is just a numbers game. There is so much more of the ANT that it is what you see.

Halestrap: What is the story with AMP? My recollection is that you get an AMP effect on proton conductance. Do you think that is through binding to the ANT?

Brand: Yes.

Halestrap: But that activates conductance. Whereas normally one doesn't think of the ANT binding AMP (at least, it doesn't translocate it), you are saying that it binds it but doesn't translocate it, and in binding it is opening up into a proton channel.

Brand: Yes, our model is that there is an ATP binding site, and fatty acids can bind nearby and alter the ANT conformation so that the ATP no longer inhibits proton conductance. AMP binds to the ATP binding site but doesn't reach and cover the bit that would be inhibitory to proton conductance; instead, it leaves it open and displaces any endogenous ATP that might be there. Therefore it activates.

Halestrap: How does this relate to ROS?

Brand: That would be quite separate. This is the ATP inhibitory site, quite separate from the ROS site.

Halestrap: So you don't think ROS is actually blocking ATP binding.

Brand: No, there doesn't seem to be an interaction. We can inhibit the ROS-activated state by adding ATP, but we can't overcome inhibition in the presence of ATP by adding more ROS.

Orrenius: I have a question relating to the comparison between the activities of the ANT and UCP. ANT is extremely dependent on cardiolipin for function. Is there a similar association of this phospholipid with UCP? At least under apoptotic conditions, cardiolipin is the phospholipid in mitochondria which is most easily oxidized.

Brand: We don't know, but that makes sense, and I suspect it does. I like the idea that UCPs are holding cardiolipin up and waving it, and when ROS react with this sentinel cardiolipin this leads to activation of UCP.

Bernardi: Have you tried comparing the concentration of atractyloside needed for inhibiting UCPs and ANT? The reason I ask is that you need way more atractylate to open the pore than is needed to inhibit the ANT, suggesting that they may act on different proteins.

Brand: We haven't done the titrations yet, but we should.

Bernardi: Do you think there could be UCPs that are uncoupled because they don't respond to regulators? You mentioned over-expressed UCPs doing the right job without being regulated.

Brand: If any overexpressed mitochondrial inner membrane protein is denatured, like scrambled egg, then it is likely to make the membrane proton permeable. Therefore you would incorrectly score this as a native uncoupling activity. Using activators and inhibitors to define native function is quite circular, I agree, but there are independent experiments, such as detergent extractability, that suggest that over-expressed UCPs are not properly folded. The only safe test is knockout of a putative UCP: if that removes uncoupling activity, then the native UCP was active without any requirement for further regulators, but this is not found for UCP1, 2 or 3.

Schon: I have a naïve question. How does the uncoupling protein make heat?

Brand: It is straightforward. All you have to do is displace a reaction from equilibrium and it will generate heat. By increasing the rate of the electron transport chain by uncoupling, all the steps move away from equilibrium, and each step releases heat.

Nicholls: Thermodynamically, you can treat the whole system as the black box. It is what goes in and what goes out that counts. UCP1 is increasing the rate of fatty acid oxidation and the rate of oxygen uptake. The rate of conversion increases, and the heat increases as a result of that. It is totally independent of mechanism. Much of the early research on brown fat attempted to identify the site of heat production. This turned out to be futile, because the sites are everywhere. It can be simplified down to a black box, and the faster things go in and go out the more heat.

Schon: Once the mechanism of ATP synthesis was elucidated, you'd imagine this mechanism would make a lot of heat, unless it is more than say 95% efficient.

Nicholls: We have calculated how many millivolts you have to drop the protonmotive force for the ATP synthase to go backwards instead of forwards. If you take a starting value of 150 mV and inhibit mitochondria in cells with rotenone, membrane potential drops to 125 mV (Nicholls 2006), so a 25 mV change is required for the switch from working rapidly forwards to working backwards.

NOVEL UNCOUPLING PROTEINS

Spiegelman: Here's another way you could think about this. You need the right amount of ATP. If you are uncoupling you are running fuel oxidation faster to maintain ATP homeostasis. By running this at a higher state and maintaining ATP homeostasis, you will generate heat. Let's say you could remove the ATP, such as with actin–myosin ATPase. You would also get heat.

Nicholls: It is totally mechanism independent.

Let's move on to this question of mild uncoupling and relief from oxidative stress. I have some problems in terms of mechanism. Much of the literature on uncoupling proteins tends to forget that uncouplers uncouple. For every bit of increased proton conductance you are inducing to try to decrease membrane potential, you are proportionately decreasing the maximal capacity of those mitochondria to generate ATP. In particular, in the neuronal context our take is that the problems from decreased ATP-generating capacity greatly outweigh any advantages from decreases in ROS production. Is the field in general taking enough account of the fact that uncouplers uncouple in terms of ATP generation?

Brand: Of course, uncoupling will reduce ATP generation. The question is, how much? The answer depends on the relative responses to the PMF of the machinery that makes ATP and the uncoupling pathway. The way to access this experimentally is with a control analysis, asking how much a little uncoupling inhibits ATP synthesis. We have done this in isolated muscle mitochondria, isolated liver mitochondria, intact hepatocytes and perfused muscle. In all these cases the answer is not much. Mild uncoupling does very little to ATP synthesis rates, but can have strong effects on ROS production in the same systems.

Spiegelman: It would be self-correcting because of the change in ADP:ATP ratio.

Nicholls: Yes, but there is a maximal respiratory rate that we call uncontrolled respiration. Let's say that is 100 units, and you start off with 20 units of proton leak, which gives 80 units of ATP production. If you increase to 30 or 40 units of leak, maximal ATP production drops to 60 to 70. The effect of mild uncoupling on ATP synthesis rates may be right in terms of control analysis. But in many tissues you are lowering your spare respiratory capacity, so in the brain, for example, you are lowering your safety margin for when those neurons have to face an epileptiform stimulation.

Spiegelman: We have a recent paper in which we titrate chemical uncoupling and find a proportional induction of PGC1α, and it returned to homeostasis. You can take mitochondria through quite a range of uncoupling and see a drop in ATP level, but this level will come back exactly to the starting point in a way that depends on PGC1α (Rohas et al 2007).

Nicholls: I find this exciting, because one thing that worried me about your model is that you said the primary inducer of PGC1α was Ca^{2+}/calmodulin. What makes more sense is if it is a drop in ATP:ADP ratio or even an increase in AMP.

Spiegelman: These all work. What is shocking to us is that we can chemically uncouple mitochondria and see a drop in ATP, but in wild-type cells it will come back precisely to the starting point. In a PGC1α knockout it drops but doesn't come back. This is a homeostatic system that can take an insult and respond to it.

Nicholls: We have worked with synaptosomes and cultured neurons. In both, the overwhelming potential use of ATP is the Na^+/K^+ ATPase in the plasma membrane. This can be short circuited by addition of a Na^+ channel activator. In both conditions the maximal activity of the mitochondria is enough to cope with a short circuited Na^+ pump plus a little bit more. It seems to me that there is always just the right safety margin of ATP-generating capacity.

Spiegelman: In the last paragraph of our paper people will think we have lost it (Rohas et al 2007): we suggest that people should reinvestigate DNP as a treatment for human obesity, in light of the compensatory systems. It works in obesity. It is toxic, but not as toxic as you might have thought. We think there is a compensatory system that gives a reasonable margin for safe use.

Lemasters: Some years ago Jorg Stucki had the idea of optimal degrees of coupling from the non-linear thermodynamic analysis he did (Stucki 1980). Degree of coupling (q) was a variable between 0 and 1, and he talked about how efficient energy transfer was in any particular system. You would think 100% efficiency ($q = 1$) would be the best, but when this occurs you cannot have any movement through the metabolic pathways. Stucki calculated degrees of coupling for maximum efficiency, power output and so on. It struck me that the uncoupling proteins would be the ideal way to adjust the degree of coupling to optimize for what might be biologically needed. Do you want maximum efficiency to survive a famine or maximum flux to escape a predator?

Brand: That's a good point. There are irreversible thermodynamic reasons why one might wish to modify efficiency. It is not as straightforward as we tend to think. There's no evidence that modulating efficiency for this purpose is or is not an important function of UCPs.

Spiegelman: If you didn't have uncoupling, wouldn't the variable nature of ATP turnover, at least in some tissues, be dangerous? For example, in skeletal muscle, you want the oxidative capacity to produce ATP to run. On the other hand, when you are not running this becomes a dangerous machine because there would be a backing up of the electron transport chain because of the lack of ATP turnover.

Nicholls: This is where the endogenous proton leak comes in. It has the advantage of being voltage dependent. The leak is high in state IV and low in state III. This goes back to Britton and Chance in 1955, and it has come full circle today. If you are doing classic oxygen electrode experiments, state IV–state III–state IV, convention says that you always take the total oxygen uptake during the stimulated respiration as the divisor for your ADP:O ratio, rather than the extra uptake. No one knew why this was. If this is interpreted in terms of work Martin and I have been doing, it means that during this state III respiration the proton leak stops and all the protons go through the ATP synthase.

Lemasters: I think that is a gross exaggeration! First, the membrane potential does not change that much during the transition from state III to state IV, and especially *in situ* in cells where membrane potential is not far out on the curve where the leak kicks in a great deal. The membrane potential changes *in vivo* in the state III to IV and state IV to III transitions are not so large as to kick in a lot of nonlinear leak.

Nicholls: Also, if the membrane potential is changing so little as respiration increases, is it really a viable mechanism for controlling ROS?

Halestrap: Then you need to remember substrate supply. There may also be regulation by increasing the activity of dehydrogenases and hence the supply of NADH to fuel the respiratory chain. This may lead to a higher membrane potential.

Bednarczyk: I have a simple question. ATP blocks UCP. Is this by phosphorylation?

Brand: For UCP1, where it is best worked out, it is thought to be non-covalent binding. It is an allosteric inhibition.

Bednarczyk: Is ADP different?

Brand: There's not much difference between ATP, ADP, GTP and GDP. It is purine nucleotides binding to a site. For UCP2 and 3 we just translocate that idea across.

Martinou: Is there any possibility that the UCP translocates ATP like ANT does? You mentioned similarities between UCPs and ANT at the beginning?

Nicholls: GDP did not get into the mitochondrial matrix, so it works but it is not transported.

Shirihai: I very much like your model that UCP2 is a signalling molecule, but not for attenuating insulin secretion—rather improving it. Recent data suggest that UCP2 knockout mice are not protected from glucotoxicity and do not show improved glucose stimulated insulin secretion. This is in contrast to previously published data from the study of a knockout mouse generated by the group of Brad Lowell (Zhang et al 2001). Lowell and colleagues have reported that glucose stimulated insulin secretion and ATP/ADP were reduced. The combination of the two strains of mice used as the embryonic stem (ES) cells and the breeder generated an inherent problem since the ES cells are derived from a good insulin secretor and the breeder is a poor secretor, leading to an unfortunate possibility that the improved secretion of insulin in the knockout mice is due to the genetic background of the ES cells. Sheila Collins gave a presentation at the ADA meeting where she described an intensive study in which her group worked to clean the genetic background of the knockout mouse. The reason this was required is that the ES cells used for targeting the gene are coming from a very good insulin secretor mouse strain and the wild-type mouse used in this study is a poor insulin secretor. After close to 20 generations of backcrossing Dr Collins observed a reversal of the difference between the wild-type and the knockout mouse, meaning that the knockout had reduced glucose stimulated insulin secretion. She then repeated this in two more strains and found similar results, UCP2 knockout compromises insulin secretion. This indicates that not far from the UCP2 locus on the mouse chromosome there is a gene that regulates insulin secretion and is likely to be the reason for the increased secretion of the UCP2 knockout from the Brad Lowell group.

Nicholls: So the knockout is not a UCP2 knockout, but a knockout of something else which is causing the phenotype?

Shirihai: It is a combination of UCP2 and another gene that is probably located nearby. I hope that Dr Collins will pursue her study on this front and identify the other gene as this one is clearly a modifier of insulin secretion.

Brand: I can't explain the inverse effect but the dependence on background is very well explained by the model. It is becoming increasingly clear that one of the main reasons for the C57BL/6 phenotype (impaired insulin secretion, propensity to obesity and diabetes) is that it's a knockout for the mitochondrial NADH/NADP transhydrogenase (NNT) that keeps mitochondrial matrix glutathione reduced and able to remove mitochondrial ROS. In C57BL/6 mice you therefore have a pre-oxidatively stressed β cell, which would activate the UCP2. Therefore the UCP2 knockout should be much more obvious in the C57BL/6 background than it is in a wild-type mouse which doesn't have the NNT knockout.

Turnbull: You think that some of the ROS is arising from fatty acid oxidation. Whereabouts? Are you are aware that there are plenty of patients who have defects of mitochondrial fatty acid oxidation, and I am not sure that they have much abnormality of insulin secretion?

Brand: It would depend where the defect in fatty acid oxidation is: whether it is upstream or downstream of the ROS-producing sites. Isolated mitochondrial work shows that the fatty acids will produce ROS at two different sites. One is by the ETFQ oxidoreductase, and the other is by either reverse electron transport or stalled complex I. If we examine fatty acid oxidation we get almost as much ROS as we do if we use succinate to reverse electron transport.

Turnbull: Under those circumstances, if patients have defects of the acyl-CoA dehydrogenases then they should be protected against ROS. This should affect insulin secretion. It would be straightforward to look at the insulin secretion of these patients.

References

Fridell Y-W, Sánchez-Blanco A, Silvia BA, Helfand SL 2004 Functional characterization of a *Drosophila* mitochondrial uncoupling protein. J Bioenerg Biomembr 36:219–228

Nicholls DG 1977 The effective proton conductances of the inner membrane of mitochondria from brown adipose tissue: dependency on proton electrochemical gradient. Eur J Biochem 77:349–356

Nicholls DG 2006 Simultaneous monitoring of ionophore- and inhibitor-mediated plasma and mitochondrial membrane potential changes in cultured neurons. J Biol Chem 281:14864–14874

Rohas LM, St-Pierre J, Uldry M, Jäger S, Handschin C, Spiegelman BM 2007 A fundamental system of cellular energy homeostasis regulated by PGC-1? Proc Natl Acad Sci USA 104:7933–7938

Stucki JW 1980 The optimal efficiency and the economic degrees of coupling of oxidative phosphorylation. Eur J Biochem 109:269–283

Zhang C-Y, Baffy G, Perret P et al 2001 Uncoupling protein-2 negatively regulates insulin secretion and is a major link between obesity, β cell dysfunction, and type 2 diabetes. Cell 105:745–55

Mitochondria as generators and targets of nitric oxide

Cecilia Giulivi

Department of Molecular Biosciences, University of California, Davis, CA 95616, USA

Abstract. Mitochondrial biochemistry is complex, expanding from oxidative phosphorylation, lipid catabolism and haem biosynthesis, to apoptosis, calcium homeostasis, and production of reactive oxygen and nitrogen species, including nitric oxide (NO). This molecule is produced by a mitochondrial nitric-oxide synthase (mtNOS). The rates of consumption and production determine the steady-state concentration of NO at subcellular levels, leading to the regulation of several mitochondrial events. Temporospatial processes tightly regulate the production of NO in mitochondria to maximize target effects and minimize deleterious reactions. Temporal regulatory mechanisms of mtNOS include activation by calcium and transcriptional/translational regulation. Calcium-activated mtNOS inhibits mitochondrial respiration. This regulation antagonizes the effects of calcium on matrix calcium-dependent dehydrogenases, preventing the formation of anoxic foci. Temporal regulation of NO production by intracellular calcium signalling requires the understanding of the heterogeneous intracellular calcium response and calcium distribution. NO production in mitochondria is spatially regulated by subcellular localization of mtNOS (e.g. acylation and protein–protein interactions), in addition to transcriptional regulation. Given the short half-life of NO in biological systems, organelle localization of mtNOS is crucial for NO to function as a signal molecule. These temporospatial processes are biologically important to allow NO to act as an effective signal molecule to regulate mitochondrial events.

2007 Mitochondrial biology: new perspectives. Wiley, Chichester (Novartis Foundation Symposium 287) p 92–104

Mitochondria are considered the main source of intracellular reactive oxygen species (ROS) production under normal, physiological conditions via complex III and complex I. The presence of a constitutive nitric oxide synthase in mitochondria (mtNOS; Kato & Giulivi 2006, see references therein) has offered the opportunity to review several aspects of mitochondrial physiology. For example, with the presence of NO at intracellular settings, it is more appropriate to explore the biological effects of reactive nitrogen and oxygen species (RNOS), including the diverse chemistry of these species. Considering that RNOS can be viewed as a part of signal transduction pathways as well as damaging species (due to their high

NITRIC OXIDE

reactivity) regulation of NO production is expected to play a critical role in determining RNOS final biological effects.

NO production in mitochondria is a tightly regulated process, controlled by temporospatial mechanisms. Calcium signalling plays a central role in temporal regulation of NO production in mitochondria given that NOS is activated by calcium. Post-translational modifications of mtNOS can be viewed as a temporal regulatory mechanism resulting in the modulation of its enzymatic activity. Spatial regulatory mechanisms of NO production allow a localized production of NO close to its targets, minimizing undesirable side reactions. Regulation of NO production will ultimately affect its interaction with target molecules, i.e. cytochrome c oxidase.

In the following sections, I will discuss various aspects of the control of production of NO in mitochondria and the biological significance of the NO/cytochrome c oxidase interaction.

Dynamic control of NO production: when and where?

In vivo, two opposite processes regulate the steady-state concentration of NO: consumption and production of NO.

The main consumption pathways of NO include its reaction with oxymyoglobin or oxyhaemoglobin (yielding nitrate) (Eich et al 1996), with superoxide anion to produce the highly reactive species peroxynitrite (Radi et al 2002), and with oxygen (Liu et al 1998). Reaction of NO with oxymyoglobin or oxyhaemoglobin serves as an important mechanism of NO consumption in the red blood cells, smooth and cardiac muscle where these haemoproteins are present at high millimolar levels. In this regard, oxymyoglobin has been proposed to serve as an NO-scavenging molecule rather than having a critical role in oxygen storage (Merx et al 2005). Peroxynitrite formation, and its subsequent reaction with biomolecules, may represent the nitrative stress component of NO effects. It has been demonstrated that peroxynitrite generation could lead to nitrative/oxidative damage to the electron transport chain and ATPase activity, altering mitochondrial ATP production (Radi et al 1994). For example, we have shown that nitrated cytochrome c is less likely to be released from mitochondria and does not initiate apoptosis via Apaf-1 (Oursler et al 2005). Although the third pathway, namely the reaction between NO and oxygen, is slow (Liu et al 1998), it may represent a critical clearance mechanism in a hydrophobic milieu, i.e., membranes (Ford et al 1993). As a result, the concentration of NO can be adequately balanced by its production and consumption, allowing NO to operate as a signalling molecule.

Regulation of NO production can be achieved through several mechanisms, such as alteration of NOS expression, modulation of NOS activity by calcium, through post-translational modifications, or protein–protein interactions.

Translational and transcriptional regulation of mtNOS expression was demonstrated to be an important mechanism of temporal regulation of cellular NO production (Kato & Giulivi 2006). For instance, inducible NOS (iNOS) is overexpressed in response to infectious pathogens through transcription regulation (Fang 2004). Considering the identification of mtNOS as an nNOS isoform (Elfering et al 2002), it could be assumed that regulatory mechanisms for nNOS expression are operative for mtNOS. For example, a complex use of nNOS alternative promoters can affect transcription patterns and translational efficiencies (Landry et al 2003). In this regard, a recent study indicated that hypoxia rapidly up-regulated nNOS expression (Ward et al 2005). These findings imply that living organisms have tightly regulated transcriptional/translational mechanisms for controlling nNOS expression in response to milieu changes. As an extension, it is possible that expression of mtNOS is tightly regulated at the transcriptional/translational steps.

Calcium serves as a major regulator of NO production. It is well established that calcium signalling plays a crucial role in modulating constitutive NOS activity such as endothelial NOS (eNOS) and neuronal NOS (nNOS). The calcium-dependent mtNOS activity becomes relevant considering the critical role of mitochondria at maintaining calcium homeostasis. Upon cell activation, mitochondria are exposed to different calcium concentrations; those close to endoplasmic reticulum (ER) calcium stores face microdomains of higher calcium concentrations through the opening of inositol triphosphate (IP3)-gated channels or ryanodine receptors (Rizzuto et al 1999, Montero et al 2000). Additionally, it has been indicated that ryanodine receptor isoform 1 is localized to the inner membrane of mitochondria, which may be one of the mechanisms for calcium influx into mitochondria (Beutner et al 2001). This, in addition to a calcium efflux antiporter different from the uniporter, illustrates the complexity of calcium homeostasis in mitochondria, and the difficulty in delineating the temporal regulatory mechanism of how calcium influences NO production and its inhibitory effect on cellular respiration. A clear understanding of the association between calcium and NO-regulated mitochondrial respiration should provide information on cytosolic and intramitochondrial calcium concentrations that affect mitochondrial motility and the activity of calcium-dependent mitochondrial dehydrogenases.

Calcium can regulate mitochondrial functions by activating key metabolic enzymes (pyruvate dehydrogenase, NAD^+-dependent isocitrate dehydrogenase, and 2-oxoglutarate dehydrogenase or α-ketoglutarate dehydrogenase) and mtNOS. The activation of calcium-dependent dehydrogenases increases oxygen consumption, whereas activation of mtNOS results in the opposite effect. Therefore, activated mtNOS antagonizes the effects of dehydrogenases on mitochondrial respiration, implying that mtNOS functions as a negative regulator for mitochondrial respiration, especially at higher calcium concentrations (Traaseth et al 2004). Owing to this effect, mitochondria can still produce enough ATP while avoiding the occurrence of anoxic foci. This system allows oxygen to diffuse to mitochondria located away from calcium microdomains, and these mitochondria (exposed to lower calcium) have the potential to produce ATP via the activation of calcium-dependent dehydrogenases without activating mtNOS.

Despite the clear evidence for the inhibitory role of NO on mitochondrial respiration and its modulation by calcium signalling, there is no clear understanding of the kinetics of calcium rises, mitochondrial motility, and regulation of NO production. Calcium may regulate the recruitment (or de-recruitment) of mitochondria to specific compartments in the cell. Mitochondria have higher motility at the resting level of cytosolic calcium, becoming negligible at higher calcium concentrations like those achieved upon stimulation (Yi et al 2004). This helps to provide the ATP required for the active re-uptake of calcium by endoplasmic reticulum (ER) calcium-ATPases (Xu et al 2004).

Temporal regulation of mitochondrial NO production could also be achieved through post-translational modifications of mtNOS. Due to the high homology of mtNOS to other known NOS isoforms it is reasonable to hypothesize that mtNOS activity could be modulated through the same post-translational mechanisms as other well-characterized NOS isoforms. A number of post-translational modifications involved in the regulation of differential NOS isoforms have been described (Brune & Lapetina 1991, Bredt et al 1992, Chen et al 1999, Fulton et al 1999). One of the major regulatory modifications observed in the NOS family of enzymes is phosphorylation. The importance of Thr^{495} and Ser^{1179} (numbering base on human endothelial NOS) phosphorylation on eNOS enzymatic activity has been demonstrated (Chen et al 1999, Fulton et al 1999). Other NOS isoforms such as nNOS have been shown to be phosphorylated by cAMP-dependent protein kinase, cGMP-dependent protein kinase, protein kinase C, and CaMK-II (Brune & Lapetina 1991, Bredt et al 1992). The finding of the C-terminus phosphorylation in mtNOS (Elfering et al 2002) may indicate the importance of these post-translational modifications in the regulation of mtNOS enzymatic activity, although the *in vivo* implications are not yet clear.

Phosphorylation is not the sole post-translational modification described for mtNOS. mtNOS has been found to be E-myristoylated (Elfering et al 2002). Based on the biological functions of acylation, E-myristoylation probably contributes to trafficking of mtNOS to mitochondria or to anchoring the enzyme to mitochondrial membranes. Affecting localization or targeting of mtNOS to specific subcellular compartments may assure a confined effect.

Protein-protein interactions are another factor in the determination of mtNOS localization at the subcellular level. Docking of mitochondria to cytoskeletal structures, such as intermediate filaments or microtubules, is important for the intracellular transport of mitochondria, which determines intracellular mitochondrial distribution (Wagner et al 2003). However, it is uncertain how docking of mitochondria may alter physiological processes, among them NO production and mitochondrial respiration.

nNOS α has a PDZ domain important for protein scaffolding in signal transduction (Brenman et al 1996, Sheng & Sala 2001). In this regard, nNOS can associate with postsynaptic density (PSD)95, PSD93 or skeletal muscle $α_1$-syntrophin via this domain (Brenman et al 1996, Sheng & Sala 2001), allowing targeting to neuronal postsynaptic densities (Brenman et al 1996). Some nNOS variants lacking PDZ domain such as nNOS β and γ may have distinctive functions owing to altered protein–protein interactions (Brenman et al 1996). A physical interaction via the PDZ domain has been demonstrated between subunit Va of cytochrome c oxidase and mtNOS, using electron microscopy and immunoprecipitation experiments. Intriguingly, the interaction between cytochrome c oxidase and mtNOS depends on calcium (V. Haynes, N. N. Sinitsyna, and C. Giulivi, unpublished observations). This implies that spatial control might also be affected by temporal control, specifically for calcium signalling.

The temporospatial regulation of NO production in mitochondria may lead to important biological events. One of the downstream outcomes is intracellular oxygen redistribution. Cells have intracellular gradients of oxygen (Takahashi et al 1998) from the plasma membrane to mitochondria, where oxygen is consumed at a high rate by coupling electron transport to oxidative phosphorylation. NO endogenously produced inhibits cytochrome c oxidase activity, decreasing oxygen consumption, thus allowing other enzymes with a higher K_m for oxygen to be active (Giulivi 2003, Haynes et al 2004).

Mechanisms of the interaction of NO with cytochrome c oxidase

Besides guanylate cyclase, another target for NO seems to be constituted by cytochrome c oxidase. The oxygen reactive site in cytochrome c oxidase contains both haem iron (a_3) and copper (Cu_B) centres. NO inhibits cytochrome c oxidase in both an oxygen competitive (at haem a_3) and oxygen independent (at Cu_B) manner. Prior to inhibition of oxygen consumption, changes can be observed in enzyme and substrate (cytochrome c redox state changes). Indeed, we have demonstrated that changes in the redox state of the carriers take place, where the cross-over point was identified as cytochrome c oxidase (Sarkela et al 2001). Physiological consequences can be mediated either by direct 'metabolic' effects on oxygen consumption and/or ATP production, or via indirect 'signalling' effects via mitochondrial redox state changes and/or free radical production. *In vivo* organ and whole body measurements of NOS inhibition suggest a possible role for NO in the inhibition of cytochrome c oxidase. However, a detailed mapping of NO and oxygen levels, combined with direct measures of cytochrome c oxidase/NO binding in *in vivo* systems has not been accomplished.

NO inhibition of cytochrome c oxidase was shown to be enhanced at low oxygen tensions. It was suggested that the inhibition was competitive, indicating that the inhibitor (NO) was binding to the 'active site' displacing oxygen (substrate) (Brown & Cooper 1994). However, more recently it has been shown that non-competitive inhibition predominates at low enzyme turnover and competitive inhibition at high turnover. The NO production rate is also likely to be oxygen concentration dependent because NOS has a higher K_m (Rengasamy & Johns 1996) for oxygen (5–20 µM) than cytochrome c oxidase (<1 µM). If this situation is extrapolated to *in vivo* conditions, then the dependency of NOS and cytochrome c oxidase on oxygen suggests that the NO/cytochrome c oxidase interaction could be *less* effective at low pO_2.

Besides the inhibition of oxygen consumption, the interaction of NO with cytochrome c oxidase will result in a decreased rate of oxidation of cytochrome c. Assuming that the rate of electron transfer from Complex I (or II) into cytochrome c will not change, then one of the first events would be an increased steady-state level of reduced cytochrome c, which would attenuate or mask the inhibition of oxygen consumption. In support of this view, cells respond to hypoxia by changing the steady-state concentration of reduced cytochrome c at a significantly higher pO_2 than that required to inhibit oxygen consumption (Wilson et al 1979). Thus, it seems reasonable to incorporate these changes when using *in vitro* and *in vivo* models.

Under *in vitro* conditions, upon addition of NOS inhibitors, an overshoot of oxygen consumption is observed, probably due to the effect of NO on the inhibition of mitochondrial oxygen consumption. However, in *in vivo* models, some laboratories have seen an effect on the oxygen consumption upon inhibition of NOS (Chen et al 2002), whereas others have reported no effect on either oxygen consumption (Hiramatsu et al 1996) or the redox state of cytochrome c oxidase in brain (Wagner et al 1997). Thus, other effects may need to be considered to explain these latter *in vivo* effects, assuming that the NOS inhibitors were reaching their targets (NOS at the specific subcellular compartment) at the required concentrations for inhibition (most of them are competitive inhibitors with L-Arg).

Nitric oxide has a K_i for guanylate cyclase of 2 nM, whereas that for cytochrome c oxidase is 60 nM. This means that if cytochrome c oxidase is inhibited by NO, then the signalling through guanylate cyclase would be 100% operative. However, if cytochrome c oxidase and guanylate cyclase are exposed to different NO concentrations, or if there are gradients of pO_2 between the tissue and cytochrome c oxidase (Wittenberg & Wittenberg 1985, Takahashi et al 1998), then the previous statement is false. Furthermore, if microheterogeneity (e.g. NOS binding to cytochrome c oxidase) along with temporospatial controls are included in this discussion, then the sole comparison of K_i becomes irrelevant.

Physiological relevance of NO / cytochrome c oxidase interactions

The ability of NO to inhibit oxygen consumption has the effect of making the oxygen gradient to cells distant from the blood vessels much shallower. The possibility of *intracellular* NO/oxygen gradients between, for example, mtNOS and cytochrome c oxidase has already been discussed. Historically, cytochrome c oxidase was suggested to be involved in oxygen sensing (Mills & Jobsis 1972). Currently, the NO/cytochrome c oxidase interactions have been suggested to be involved in a range of signalling pathways in oxygen sensing. The link between the primary event (NO binding to cytochrome c oxidase) and the initiation of the signal transduction pathway could be through the modulation of mitochondrial superoxide anion production, and hence changing hydrogen peroxide concentrations in the cell. Alternatively, the decreased oxygen consumption could be related to a decreased ATP production, which would lead to changes in the energy charge of the cell (allosteric effects through AMP or activation of AMP-dependent protein kinase).

Finally, the contribution of direct biochemical effects (mediated through changes in oxygen consumption rates) and indirect biochemical effects (via changes in mitochondrial free radical production and/or ATP production) should be carefully explained, especially when extrapolations to whole organisms are sought.

Acknowledgements

This study was supported by the National Institutes of Health (ES012691 and ES005707).

References

Beutner G, Sharma VK, Giovannucci DR, Yule DI, Sheu SS 2001 Identification of a ryanodine receptor in rat heart mitochondria. J Biol Chem 276:21482–21488

Bredt DS, Ferris CD, Snyder SH 1992 Nitric oxide synthase regulatory sites. Phosphorylation by cyclic AMP-dependent protein kinase, protein kinase C, and calcium/calmodulin protein kinase; identification of flavin and calmodulin binding sites. J Biol Chem 267:10976–10981

Brenman JE, Chao DS, Gee SH et al 1996 Interaction of nitric oxide synthase with the postsynaptic density protein PSD-95 and alpha1-syntrophin mediated by PDZ domains. Cell 84:757–767

Brown GC, Cooper CE 1994 Nanomolar concentrations of nitric oxide reversibly inhibit synaptosomal respiration by competing with oxygen at cytochrome oxidase. FEBS Lett 356:295–298

Brune B, Lapetina EG 1991 Phosphorylation of nitric oxide synthase by protein kinase A. Biochem Biophys Res Commun 181:921–926

Chen Y, Traverse JH, Du R, Hou M, Bache RJ 2002 Nitric oxide modulates myocardial oxygen consumption in the failing heart. Circulation 106:273–279

Chen ZP, Mitchelhill KI, Michell BJ et al 1999 AMP-activated protein kinase phosphorylation of endothelial NO synthase. FEBS Lett 443:285–289

Eich RF, Li T, Lemon DD et al 1996 Mechanism of NO-induced oxidation of myoglobin and hemoglobin. Biochemistry 35:6976–6983

Elfering SL, Sarkela TM, Giulivi C 2002 Biochemistry of mitochondrial nitric-oxide synthase. J Biol Chem 277:38079–38086

Fang FC 2004 Antimicrobial reactive oxygen and nitrogen species: concepts and controversies. Nat Rev Microbiol 2:820–832

Ford PC, Wink DA, Stanbury DM 1993 Autoxidation kinetics of aqueous nitric oxide. FEBS Lett 326:1–3

Fulton D, Gratton JP, McCabe TJ et al 1999 Regulation of endothelium-derived nitric oxide production by the protein kinase Akt. Nature 399:597–601

Giulivi C 2003 Characterization and function of mitochondrial nitric-oxide synthase. Free Radic Biol Med 34:397–408

Haynes V, Elfering S, Traaseth N, Giulivi C 2004 Mitochondrial nitric-oxide synthase: enzyme expression, characterization, and regulation. J Bioenerg Biomembr 36:341–346

Hiramatsu T, Jonas RA, Miura T et al 1996 Cerebral metabolic recovery from deep hypothermic circulatory arrest after treatment with arginine and nitro-arginine methyl ester. J Thorac Cardiovasc Surg 112:698–707

Kato K, Giulivi C 2006 Critical overview of mitochondrial nitric-oxide synthase. Front Biosci 11:2725–2738

Landry JR, Mager DL, Wilhelm BT 2003 Complex controls: the role of alternative promoters in mammalian genomes. Trends Genet 19:640–648

Liu X, Miller MJ, Joshi MS, Thomas DD, Lancaster JR, Jr. 1998 Accelerated reaction of nitric oxide with O2 within the hydrophobic interior of biological membranes. Proc Natl Acad Sci USA 95:2175–2179

Merx MW, Godecke A, Flogel U, Schrader J 2005 Oxygen supply and nitric oxide scavenging by myoglobin contribute to exercise endurance and cardiac function. FASEB J 19:1015–1017

Mills E, Jobsis FF 1972 Mitochondrial respiratory chain of carotid body and chemoreceptor response to changes in oxygen tension. J Neurophysiol 35:405–428

Montero M, Alonso MT, Carnicero E et al 2000 Chromaffin-cell stimulation triggers fast millimolar mitochondrial Ca2+ transients that modulate secretion. Nat Cell Biol 2:57–61

Oursler MJ, Bradley EW, Elfering SL, Giulivi C 2005 Native, not nitrated, cytochrome c and mitochondria-derived hydrogen peroxide drive osteoclast apoptosis. Am J Physiol Cell Physiol 288:C156–168

Radi R, Rodriguez M, Castro L, Telleri R 1994 Inhibition of mitochondrial electron transport by peroxynitrite. Arch Biochem Biophys 308:89–95

Radi R, Cassina A, Hodara R, Quijano C, Castro L 2002 Peroxynitrite reactions and formation in mitochondria. Free Radic Biol Med 33:1451–1464

Rengasamy A, Johns RA 1996 Determination of Km for oxygen of nitric oxide synthase isoforms. J Pharmacol Exp Ther 276:30–33

Rizzuto R, Pinton P, Brini M, Chiesa A, Filippin L, Pozzan T 1999 Mitochondria as biosensors of calcium microdomains. Cell Calcium 26:193–199

Sarkela TM, Berthiaume J, Elfering S, Gybina AA, Giulivi C 2001 The modulation of oxygen radical production by nitric oxide in mitochondria. J Biol Chem 276:6945–6949

Sheng M, Sala C 2001 PDZ domains and the organization of supramolecular complexes. Annu Rev Neurosci 24:1–29

Takahashi E, Sato K, Endoh H, Xu ZL, Doi K 1998 Direct observation of radial intracellular PO2 gradients in a single cardiomyocyte of the rat. Am J Physiol 275:H225–233

Traaseth N, Elfering S, Solien J, Haynes V, Giulivi C 2004 Role of calcium signaling in the activation of mitochondrial nitric oxide synthase and citric acid cycle. Biochim Biophys Acta 1658:64–71

Wagner BP, Stingele R, Williams MA, Wilson DA, Traystman RJ, Hanley DF 1997 NO contributes to neurohypophysial but not other regional cerebral fluorocarbon-induced hyperemia in cats. Am J Physiol 273:H1994–2000

Wagner OI, Lifshitz J, Janmey PA, Linden M, McIntosh TK, Leterrier JF 2003 Mechanisms of mitochondria-neurofilament interactions. J Neurosci 23:9046–9058

Ward ME, Toporsian M, Scott JA et al 2005 Hypoxia induces a functionally significant and translationally efficient neuronal NO synthase mRNA variant. J Clin Invest 115:3128–3139

Wilson DF, Erecinska M, Drown C, Silver IA 1979 The oxygen dependence of cellular energy metabolism. Arch Biochem Biophys 195:485–493

Wittenberg BA, Wittenberg JB 1985 Oxygen pressure gradients in isolated cardiac myocytes. J Biol Chem 260:6548–6554

Xu W, Liu L, Charles IG, Moncada S 2004 Nitric oxide induces coupling of mitochondrial signalling with the endoplasmic reticulum stress response. Nat Cell Biol 6:1129–1134

Yi M, Weaver D, Hajnoczky G 2004 Control of mitochondrial motility and distribution by the calcium signal: a homeostatic circuit. J Cell Biol 167:661–672

DISCUSSION

Youle: Where is nNOS expressed in the body?

Giulivi: The nomenclature is a problem. nNOS was first found in neurons, but we have found it in every tissue.

Youle: Is the same splice variant expressed outside and inside the mitochondria?

Giulivi: It depends on which brain region you are talking about. Sometimes you can get coexpression of NOSµ and NOSα.

Youle: Let's say the mitochondrial isoform can also be expressed outside the mitochondria. How would this targeting be specified?

Giulivi: My guess is that the PDZ domain plays a fundamental role. In brain it has to be bound to dystrophin, for example, which gives the localization.

Schon: The demonstration that NOS is inside mitochondria is critical. This needs to be shown using standard experimental approaches. This molecule doesn't have an obvious targeting signal. It must have some targeting information but it is clearly not in the N-terminus like a standard targeting signal, so for that reason one has to do a little gene bashing with a marker protein to show that it is going in, or do an importation study, even in yeast. I haven't seen any western blots showing that it is mitochondrial.

Giulivi: We did import experiments. We did *in vitro* translation. It is ATP driven (sensitive to FCCP), and the product is resistant to trypsin treatment, and gets inside within a minute. Several mitochondrial proteins get targeted to mitochondria without having an obvious N-terminal targeting sequence.

Schon: Is it FCCP sensitive?

Giulivi: Yes. We did the best that we could do. Then we also used mitoplasts to avoid the issue of outer membrane, and we did two-dimensional gel electrophoresis of the 'precursor' and 'mature' proteins. Both showed identical pI and MW.

Schon: NO is freely diffusible across membranes. The question is, do you need to compartmentalize nNOS inside the matrix when NO is a famous intercellular second messenger? Are there gradients of NO concentration within a single cell?

Giulivi: When the activity of NOS in intact mitochondria is measured, we get a rate of anything from 2 and 2.5 nmoles/min per mg protein. If we put an electrode in very little is coming out: just 10–20% of the amount produced. The compartmentalization issue is better understood when we consider the solubility of NO (it is as hydrophobic as oxygen, and we know that there are oxygen gradients in cells), and the way that NOSs exist in biological systems (except for iNOS, all others interact with or bind other proteins, e.g. PSD60, PSD90, caveolin, etc.).

Rizzuto: I share Richard Youle's concern. We teach students that nNOS is neuronal. You can find some in other cell types, but it is mainly neuronal, if you look simply at expression levels.

Giulivi: We are talking about something else now. There are different roles for the same molecule in different tissues. In neurons it might have a primary role as a neurotransmitter. In mitochondria it has a different role.

Rizzuto: If we accept that there is a lot in neurons and much less in other tissues, and there is no splice variant difference, we are talking about exactly the same molecule. This means that there is only a fraction in neurons and presumably the same fraction in other cells that gets into mitochondria unless something specific happens. Did you calculate how much is proteinase K resistant when you isolate mitochondria, to make sure that it is really in the matrix? If the neuronal proportion holds true, I assume it will be a small fraction compared with the signalling element present in the cytoplasm. If it is just a small fraction of a protein that is expressed at low levels in non-neuronal cells, how much nNOS is really mitochondrial?

Giulivi: Which partition are you talking about? Let's talk about liver. In liver we find 90% of nNOS in mitochondria. In neurons it is different because its main role is as a neurotransmitter, so we didn't look at this.

Rizzuto: Is it proteinase K resistant?

Giulivi: Yes.

Spiegelman: There may be no general answer to this, but when peroxynitrite is generated, to what extent does this occur as a product from the generation of NO within the mitochondria? Does it matter in terms of peroxynitrite?

Giulivi: There are two views on this. The rate constant of the superoxide ion with SOD is $2 \times 10^9 \, M^{-1} \, s^{-1}$. There is micromolar SOD in the matrix that is way above the concentration of the substrate. The main thing will be driven by the catabolism of SOD. A small amount will be driven by NO and superoxide ion. We are talking about physical chemistry in a homogeneous system. In mitochondria, the matrix has 80–90 mg protein per ml. At this protein concentration, the matrix looks like a gel more than a homogeneous aqueous system, thus the rate constants from physical chemistry may not apply. In addition, a different chemistry between NO and oxygen is taking place at the membrane, again driven by their high concentration in this compartment. Others have reported that increasing NO production results in more peroxynitrite and more nitration, but this makes me think that there must be some compartmentalization (as we published before on the nitration of mitochondrial proteins), or that SOD is not catabolizing all the superoxide ion.

Spiegelman: If I could make a suggestion, I would find the mitochondrial targeting sequence and mutate it *in vivo*. I would make an nNOS allele that cannot go into the mitochondrion. This would shut everyone up!

Giulivi: There are several proteins that get into mitochondria that don't have a targeting sequence.

Spiegelman: There must be a mechanism. You will find signals that allow you to distinguish at some level. This is the only way you'll be allowed to separate these compartments in a way that is totally convincing.

Giulivi: We started looking at contact sites. We did *in vitro* translation and IP of isolated outer membrane to see what proteins are coming down with this NOS.

Spiegelman: You don't even need to know that. You can work at the level of *cis*-acting sequences and primary sequence. Who they play with is interesting, but you don't need to know this. What you need is to find a mutant or a place to attack.

Orrenius: I am thinking about the localization of potential NO targets, and what is available and not available to the generated NO. We know that superoxide generated by the respiratory chain may be released either into the matrix or into the intermembrane space. Would superoxide in the intermembrane space also be available to react with NO? Further, there is a possibility that regulation of respiratory activity may not only occur by interaction with cytochrome oxidase but also with cytochrome c. NO can modulate cytochrome c activity by nitrosylation.

Giulivi: There are two questions there. First, is there no other level to modulate superoxide ion in the intermembrane space? The answer is the same as I gave before for the matrix SOD. There is a copper/zinc SOD in the intermembrane space. It is possible, but I don't know how much. The second question was the nitration of cytochrome c. In our hands this causes it to lose its capacity to leave mitochondria and initiate apoptosis. I am not sure that this is a regulatory mechanism; it's probably a dead-end pathway.

Rich: I was interested in your comment that the enzyme binds to subunit 5A of cytochrome oxidase. What is the evidence for this? Its binding to the matrix domain of subunit 5A would not put it in a very useful location in terms of targeting its product to cytochrome oxidase since NO enters the active site of cytochrome oxidase through a membrane-buried channel in subunit 3.

Giulivi: We didn't do the 5A binding studies. These were done by Persichini and colleagues using IP of NOS in intact cells. They looked at what came down. In our case, we followed a different approach. We found that binding to 5A is secondary to the binding of another cytochrome oxidase subunit. If NOS is not bound to this particular subunit, it does not bind to 5A. This is what we are finding. One of the main criticisms of this study was that 5A is too far out of the active site.

Nicholls: I am slightly concerned about a mechanism that inhibits oxygen availability in terms of what physiological effects this is going to have.

Duchen: The idea that NO is helping to define oxygen gradients from the endothelium is a nice one, but it means that the NO is having to be most effective where oxygen is highest, and yet it is a competitive antagonist for oxygen. This means you need to make a lot of NO in the endothelial cells to combat the locally high oxygen tension.

Giulivi: I don't understand why you'd need to produce a lot of NO.

Nicholls: The idea is that NO is trying to even out the gradient, but you are saying that it has to be highly effective near the blood vessels where the oxygen concentration is the highest and therefore the competition is most difficult.

Giulivi: The amount of NO is in the low nanomolar range. For oxygen it is higher than that. Very little NO is needed to produce inhibition. Does this answer your question?

Duchen: Not really. It seems counterintuitive to have a competitive antagonist that is having to compete with oxygen at a place where the oxygen concentration is highest.

Giulivi: How do you put oxygen back where you don't have enough, unless the ones that are consuming the most share their wealth with the others at the back?

Lemasters: The idea that some cells are further away from blood vessels than others doesn't hold in the liver and heart. Every parenchymal cell is next to one or more capillaries. You might have gradients of individual mitochondria being closer to the oxygen source than others, but a single hepatocyte might be bordering three different sinusoids. The same is true for cardiac myocytes: the heart is very well vascularized.

Giulivi: We have some preliminary results on this issue. We found (as have others) gradients of oxygen in the liver. We did laser capture microscopy for hepatocytes close to the arteries versus those close to veins. We did western blots to look for NOS. Those close to the artery had more NOS than those close to the vein.

O'Rourke: How much of the mitochondrial NO that is generated is coming from NOS, as opposed to other sources. Jay Zweier's group has reported peroxynitrite production at complex III by the reaction of NO with superoxide (Chen et al

2006). He has done this by EPR measurements on mitochondria. Can you discriminate between these two sources? The alternative NO source I was thinking of was from complex IV, as described in Castello et al (2006).

Giulivi: By EPR measurements, the major source is NOS. There is another source we detected by EPR not inhibitable by NOS inhibitors. This source is about 10% of the other production. Since mitochondria have GSNO, we attributed this slow, non-enzymatic production to GSNO decay.

Spiegelman: Given that reactive nitrogen species are probably inevitable, can you comment on the mechanisms for detoxification?

Giulivi: For peroxynitrite it is complicated, because it is like trapping hydroxyl radicals. It is very reactive, and the radius of reaction is narrow. Usually it is in the anionic form, so we need an antioxidant that goes to the soluble, aqueous form. Some groups have published that peroxynitrite can cross the membrane in the protonated form. In terms of detoxification, in the case of nitration of proteins, this seems to be a dead-end road, and they get catabolized by the proteolytic pathway. There are no enzymatic steps for detoxification.

References

Chen YR, Chen CL, Yeh A, Liu X, Zweier JL 2006 Direct and indirect roles of cytochrome b in the mediation of superoxide generation and NO catabolism by mitochondrial succinate-cytochrome c reductase. J Biol Chem 281:13159–13168

Castello PR, David PS, McClure T, Crook Z, Poyton RO 2006 Mitochondrial cytochrome oxidase produces nitric oxide under hypoxic conditions: implications for oxygen sensing and hypoxic signaling in eukaryotes. Cell Metab 3:277–287

Calcium signalling and mitochondrial motility

György Hajnóczky, Masao Saotome, György Csordás, David Weaver and Muqing Yi

Department of Pathology, Anatomy and Cell Biology, Thomas Jefferson University, Philadelphia, PA 19107, USA

Abstract. Intracellular Ca^{2+} is able to control numerous cellular responses through complex spatial and temporal organization. For the effective handling of intracellular Ca^{2+}, endoplasmic reticulum (ER) Ca^{2+} mobilization and plasma membrane Ca^{2+} entry have to be complemented by strategic and dynamic positioning of an energy source that is usually provided by mitochondrial ATP production. Mitochondria also participate in the transport of Ca^{2+}. Mitochondria are dynamically distributed in cells and utilize cytoskeletal tracks and motor proteins for their movements. Recent studies have reported that Ca^{2+} inhibits mitochondrial motility providing a mechanism to retain mitochondria at Ca^{2+} signalling sites. Here we discuss the control of the mitochondrial distribution by cell signalling mechanisms, the spatial relationship among individual mitochondria and ER domains, and the possible implications of mitochondrial movements in the Ca^{2+}-dependent cell survival and cell death mechanisms.

2007 Mitochondrial biology: new perspectives. Wiley, Chichester (Novartis Foundation Symposium 287) p 105–121

An important factor for the complex, heterogeneous and dynamic mitochondrial morphology in cells is the motion of the mitochondria. Mitochondrial movement ceases on cellular ATP depletion, indicating the active nature of the motility. Mitochondria use the cytoskeletal tracks and motor proteins to exercise their movements. Recent studies have shown that mitochondrial motility is affected by intracellular messengers, including the Ca^{2+} signal. However, the molecular mechanisms underlying the specificity and the spatiotemporal control of the mitochondrial movements by cell signalling mechanisms remain to be solved.

Redistribution of mitochondria to specific subcellular regions provides a mechanism to match the local ATP requirements. Furthermore, recruitment of mitochondria may also lead to an increase in the local cytoplasmic Ca^{2+} buffering at the sites of the Ca^{2+} release from the endoplasmic reticulum (ER) or Ca^{2+} entry at the plasma membrane. From the mitochondrial perspective, accumulation of Ca^{2+} to the mitochondrial matrix is important for the regulation of the Krebs cycle and for the control of the mitochondrial membrane permeability. Since mitochondrial movements will determine the relative position of the individual organelles, the motility is likely to affect the intermitochondrial interactions, either membraneous or chemical or electrical coupling. Due to its broad significance, the study of motility is one of the rising areas of mitochondrial biology and is likely to be of great relevance for other emerging areas highlighted in this book on mitochondrial biology. This paper describes some recent data on the mechanism of the dynamic positioning of the mitochondria and discusses how cell signalling and functions may depend on mitochondrial motility.

Mitochondrial distribution and motility

Mitochondrial morphology and distribution

Mitochondria appear in a variety of shapes from small globular structures to complex networks of long tubules (Collins & Bootman 2003, Rizzuto et al 1998). In the mitochondrial population of a cell, the ratio of the discrete particles to the interconnected complexes is determined primarily by the mitochondrial fusion and fission activities. Usually, mitochondria are widely dispersed in the cytoplasm with relatively high densities around the nucleus (Fig. 1A). Grouping of mitochondria at specific subcellular locations (e.g. perigranular area) occurs and causes functional specialization in pancreatic acinar cells (Park et al 2001). In the cytoarchitecture of the muscle, the space for the mitochondria is precisely defined and stable, resulting in a regular mitochondrial spatial pattern (Rudolf et al 2004). In the processes of neurons and other cell types, a plasma membrane-bordered narrow space is available for the mitochondria that are deposited in a row often with

FIG. 1. (A) Spatial relationship between microfilaments, microtubules and mitochondria Confocal images of an H9c2 cell coexpressing actin-GFP with mitoDsRed (*left*) and tubulin-GFP with mitoDsRed (*right*). (B) Control of mitochondrial motility by the $[Ca^{2+}]_c$ signal. (*Left*) Simultaneous measurements of mitochondrial motility and $[Ca^{2+}]_c$ in an H9c2 cell expressing mitoYFP and loaded with Fura2. The decrease in mitochondrial motility and $[Ca^{2+}]_c$ are plotted in light grey and dark grey, respectively. The cell was exposed to VP (0.6 nM) in a pulsatile manner. (*Right*) Dose–response relationship between $[Ca^{2+}]_c$ and mitochondrial motility. For quantitative estimation of the $[Ca^{2+}]_c$ dependence of the inhibition of mitochondrial motility, $[Ca^{2+}]_c$ and motility were measured in cells that were incubated in a Ca^{2+} free buffer supplemented with EGTA (2 mM) and ionomycin (10 μM) to ensure rapid equilibration of the cytosol with the extracellular $[Ca^{2+}]$, and then varying amounts of $CaCl_2$ (0–15 mM) were added. The effect of various concentrations of VP on $[Ca^{2+}]_c$ and mitochondrial motility are also shown. (C) The putative mechanism for the $[Ca^{2+}]_c$ signal-dependent control of mitochondrial motility is shown. Panels B right and C are reproduced with modification from *The Journal of Cell Biology* 2004, 167:661–672 Copyright 2004 The Rockefeller University Press.

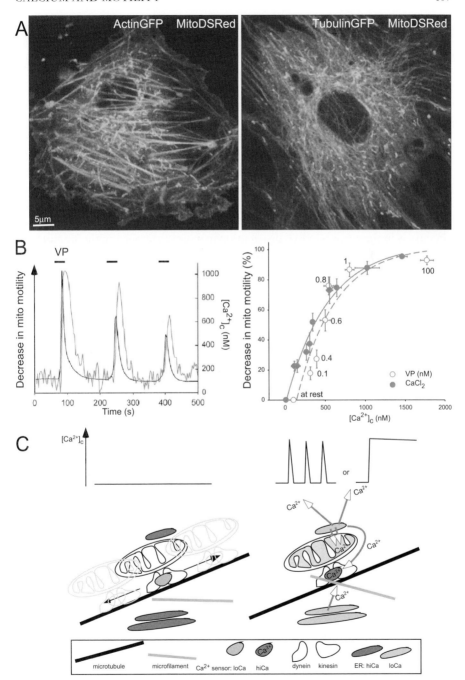

regular spacings among them. In many cell types, mitochondria seem to be scattered in the cytoplasm. However, closer observation reveals arrangements of mitochondria in tracks that follow the direction of cytoskeletal fibres (Fig. 1A).

Cytoskeletal support for the mitochondrial distribution

Positioning of mitochondria utilizes microfilaments and microtubules as well as intermediate filaments (Fig. 1A). The involvement of the various cytoskeletal components shows species, cell type and subcellular region-dependent differences. In budding yeast, mitochondria are positioned along the microfilaments and a similar arrangement occurs in the axon of neurons in mammals. However, in fission yeast and in many mammalian cells, mitochondria are commonly associated and aligned with microtubules (Hollenbeck & Saxton 2005). The desmin knockout also shows alterations in the mitochondrial distribution in muscle, supporting the significance of the intermediate filaments in mitochondrial positioning (Milner et al 2000).

Association of the mitochondria with the cytoskeletal elements is promoted by multiple mechanisms. Actin and tubulin, the building blocks of the microfilaments and microtubules, have been shown to directly interact with the voltage-dependent anion channel (VDAC) in the outer mitochondrial membrane (OMM) (Carre et al 2002, Xu et al 2001). Microtubule and microfilament associated proteins have also been involved in the binding of mitochondria to the cytoskeleton. For example, both gelsolin, a Ca^{2+}-dependent actin regulatory protein and microtubule associated protein 2 (MAP2) have been claimed to bind to the VDAC and to support the linkage of mitochondria to the cytoskeleton (Linden & Karlsson 1996, Kusano et al 2000).

Binding of molecular motors to the mitochondria provides a means for the organelles to move along the cytoskeletal fibres. Myosin motors are used for bi-directional transport of mitochondria along microfilaments, whereas conventional kinesin (kinesin 1) and cytoplasmic dynein motors facilitate the anterograde and retrograde movements along the microtubules, respectively (Vale 2003, Hollenbeck & Saxton 2005). For the binding of cytoskeletal motors, docking proteins have been identified on the mitochondria e.g. dynactin for the microtubular motor protein, cytoplasmic dynein (Habermann et al 2001, Varadi et al 2004) or for both dynein and kinesin (Deacon et al 2003). Milton, a mitochondrial protein in *Drosophila* indirectly interacts with kinesin 1, is required for localization of mitochondria to the nerve terminals (Stowers et al 2002) and shares ~44% amino acid homology with the mammalian $GABA_A$ receptor β_2 subunit-interacting, trafficking protein (GRIF1) (Brickley et al 2005). GRIF1 and the structurally homologous *O*-linked *N*-acetylglucosamine transferase interacting protein 106 (OIP106) also associate with kinesin and mitochondria in mammalian cells (Brickley et al 2005). Recently, miro, an integral OMM Rho GTPase has been revealed to interact with GRIF1 and OIP106 (Fransson et al 2006). Furthermore, miro, milton and kinesin were shown to work together in anterograde mitochondrial transport (Glater et al 2006). However, the list of the components of the motor binding complexes is likely to be far from complete and the mechanism of the anchorage remains elusive.

Mitochondrial motions

Knockout models for Kif1b, the kinesin motor that binds to the mitochondria (Tanaka et al 1998) or for the adaptor proteins (milton and miro) (Stowers et al 2002, Glater et al 2006) are embryonic lethal and mitochondria are clustered around the nucleus. The transport of mitochondria is particularly relevant in neurons because of their extended processes and high metabolic demands. Disruption of mitochondrial transport has been correlated with neurodegenerative disease (Hollenbeck & Saxton 2005). Notably, cells lacking various factors considered to be relevant for mitochondrial or ER membrane dynamics also display perinuclear mitochondrial clustering, raising the possibility that these factors could also impact on the mitochondrial transport.

Mitochondrial motility appears in the form of both long-distance travel and complex local movements. To provide the energy for the movements the motor proteins hydrolyse ATP. The average step size has been measured to be approximately 8 nm for both kinesin and dynein motors. *In vivo*, multiple kinesins or multiple dyneins seem to work together, enabling high speed movement of the organelles and the two motors of opposing polarity do not appear to work against each other (Kural et al 2005). Furthermore, both actin- and tubulin-based motors are likely to participate in the generation of complex patterns of local and long distance mitochondrial transport. Spatially and temporally co-ordinated recruitment of specific motors to the mitochondria or binding of some regulatory factors to the mitochondria-bound motors are among the mechanisms that may produce complex mitochondrial motion patterns.

Control of mitochondrial motility by the calcium signal and by other second messengers

Inhibition of mitochondrial movement has been documented at the onset of TNFα-induced cell death and was attributed to the phosphorylation of the kinesin light chain (De Vos et al 2000). This effect may be due to GSK3-mediated phosphorylation of the myosin light chain (Morfini et al 2002). Chada & Hollenbeck (2004) showed that mitochondria accumulate at local sites of nerve growth factor application, utilizing docking interactions between actin and the mitochondria. Lysophosphatidic acid has also been reported to inhibit fast mitochondrial movements, acting through the small GTPase RhoA that promotes formin-mediated binding of the organelles to the actin filaments (Minin et al 2006). In parallel, several groups have shown inhibition of mitochondrial motility in response to various Ca^{2+}-dependent stimuli in a range of cell types (Rintoul et al 2003, Yi et al 2004, Brough et al 2005). Rintoul, Reynolds and co-workers have shown that addition of glutamate to primary cultures of rat forebrain results in a rapid diminution of mitochondrial movement and also an alteration from elongated to rounded morphology. This effect required the entry of Ca^{2+} and was mediated by activation of the NMDA subtype of the glutamate receptor (Rintoul et al 2003). Our group showed that mitochondrial movement is effectively stopped during stimulation with agonists evoking either inositol-1,4,5-trisphosphate (IP_3) receptor- or ryanodine receptor-mediated cytoplasmic $[Ca^{2+}]$ ($[Ca^{2+}]_c$) spikes and oscillations in H9c2 myoblasts and in HepG2 hepatic cells (Fig. 1B *left*). Isolated spikes of mitochondrial movement inhibition were observed during low frequency $[Ca^{2+}]_c$ oscillations. However, if the frequency of $[Ca^{2+}]_c$ spiking was higher and the recovery of motility was slow, $[Ca^{2+}]_c$ oscillations could produce an essentially sustained maximal inhibition in movement activity. Thus the frequency-modulated $[Ca^{2+}]_c$ signals were translated into a time-averaged motility response (Fig. 1C) (Yi et al 2004). Brough et al showed carbachol-induced mitigation of both mitochondrial and ER motility in HEK-293 cells (Brough et al 2005). Collectively, these data have revealed that activation of both Ca^{2+} entry and intracellular Ca^{2+} mobilization leads to suppression of mitochondrial motility in a range of cell types. The possibility of cell type dependent differences in this pathway is indicated by a recent report showing that transient increases in $[Ca^{2+}]$ associated with spontaneous firing of action potentials or evoked by stimulation with a variety of agonists did not affect the mitochondrial motility in rat cortical neurons (Beltran-Parrazal et al 2006).

Ca^{2+}-dependent control of the mitochondrial movements

In H9c2 myoblasts, the inhibitory effect of the Ca^{2+}-mobilizing agonist on mitochondrial motility was reproduced by Ca^{2+} ionophore-induced increases in $[Ca^{2+}]_c$ and was in proportion to the $[Ca^{2+}]_c$ signal (Fig. 1B *right*). Also, the kinetics of the motility inhibition closely followed the time course of the $[Ca^{2+}]_c$ rise (Fig. 1B *left*). Thus, inhibition of the mitochondrial movement during Ca^{2+} mobilization can be attributed primarily to the $[Ca^{2+}]_c$ signal. Importantly, the majority of the Ca^{2+}-induced attenuation of mitochondrial motility was obtained in the submicromolar range of $[Ca^{2+}]_c$ ($IC_{50} \approx 400\,nM$), indicating that mitochondrial motility is controlled in the physiological range of $[Ca^{2+}]_c$ (Fig. 1B *right*) (Yi et al 2004). In addition, organellar motility may also be affected by metabolic changes that occur during the Ca^{2+} signal, such as ATP depletion (Brough et al 2005, Mironov 2006).

Resolution of the molecular structure did not reveal any Ca^{2+} or calmodulin binding sites in the mammalian microtubular motor proteins that support the mitochondrial movement (cytoplasmic dynein and conventional kinesin) (Vale 2003). Furthermore, the Ca^{2+} signal is not likely to control mitochondrial movement through phosphorylation or dephosphorylation of dynein or kinesin since neither KN-62, an inhibitor of Ca^{2+}/calmodulin-dependent kinases nor cyclosporin A, an inhibitor of calcineurin, the Ca^{2+}-dependent protein phosphatase, prevented the Ca^{2+}-dependent inhibition of mitochondrial motility in H9c2 myoblasts (Yi et al 2004). Thus, it seems that a distinct Ca^{2+} sensor molecule is required to translate the Ca^{2+} signal for the microtubular motor proteins (Fig. 1C).

A putative Ca^{2+} sensor is the miro that contains two EF-hands that can potentially bind Ca^{2+} (Fransson et al 2003, Frederick et al 2004). Ca^{2+} binding may control the conformation of miro, regulating its ability to recruit milton or organize the milton (GRIF-1, OIP106)–kinesin complex at the surface of mitochondria (Rice & Gelfand 2006). Another candidate for the Ca^{2+} sensor is myosin Va, which contains calmodulin binding sites and displays a Ca^{2+}-dependent interaction with actin-filaments (Tauhata et al 2001, Krementsov et al 2004) and microtubules (Cao et al 2004). Although myosin Va has been considered primarily as an actin based motor, recent studies have raised the possibility that myosin Va can interact directly with dynein and kinesin and through this interaction may affect motor function at the microtubules (Stafford et al 2000, Lalli et al 2003). Immunochemistry has revealed the presence of myosin Va on the mitochondria in some cell types (Nascimento et al 1997, Yi et al 2004) but not in others (Varadi et al 2005). The Ca^{2+}-dependent change in motility may reflect detachment of the motors from the microtubules or locking the microtubule-bound motors in a stationary position. The microtubules bind several proteins in a Ca^{2+}-dependent manner, including MAP2, MAP4 and tau (Yamamoto et al 1985), which may block indirectly by steric hindrance or by competition. However, an argument against the motor detachment is that no change in the close association between mitochondria and microtubules appeared during the period of reduced mobility (Yi et al 2004). Furthermore, after the decay of the Ca^{2+} signal, mitochondria often continue their motion without a change in the direction of motility. It is also possible that the Ca^{2+} sensor promotes anchorage of the mitochondria. For example, Ca^{2+} promotes the interaction of myosin Va with actin (Tauhata et al 2001), and linking the mitochondria to microfilaments may also inhibit the motility along microtubules (Fig. 1C).

Similar to Ca^{2+}, Zn^{2+} has also been shown to inhibit mitochondrial motions. However, the effect of Zn^{2+} on mitochondrial motility is likely to be mediated by a different mechanism through the PI3K signalling pathway (Malaiyandi et al 2005).

Motility control of interorganellar interactions

A range of signalling mechanisms and cell functions are arranged through local homotypic and heterotypic interactions between discrete organelles by means of membrane–membrane transactions, by chemical or electrical coupling. All these processes are dependent on co-ordination of the cytoskeletal and interorganellar anchorage and movements of the participating organelles. The significance of the ER–mitochondrial interplay has been exposed for example in Ca^{2+} signalling, in cell survival mechanisms and in biosynthetic processes, and the significance of the interaction between discrete mitochondria has been recognized in mitochondrial dynamics and signalling. The present account is focused on these two paradigms but the interactions of mitochondria with other structures such as plasma membrane has also been noticed and will be an important subject for future investigations (Hoth et al 2000, Parekh & Putney 2005).

ER–mitochondrial coupling

Both ER and mitochondria are anchored to the cytoskeleton and move along cytoskeletal tracks. In the perinuclear zone of the cells, the copious ER and mitochondria often form continuous networks with limited long range movements. In the peripheral regions and particularly in cellular processes discrete ER and mitochondrial domains appear, which commonly form ER–mitochondrial complexes and move in a co-ordinated manner.

Images of both fixed and live cells show tight associations between subsets of mitochondria and subdomains of the ER (Shore & Tata 1977, Rizzuto et al 1998). The cytoskeletal support of the organelles may provide conditions for the associations. However, at the ER–mitochondrial interface, tethering structures have also been visualized, which appear to be protein bridges between the ER membrane and the OMM (Mannella 2003, Csordas et al 2006). Both the intact tethers and the mitochondrial motility inhibition are relevant for the Ca^{2+} signal propagation from ER to the mitochondria (Csordas et al 2006, Yi et al 2004). These two factors might affect distinct subsets of mitochondria or work together to co-ordinate the alignment of the ER and OMM Ca^{2+} transport sites. Some evidence suggests that the Ca^{2+} transport sites show non-homogeneous membrane distribution and form clusters in the ER and OMM that are aligned with each other at the ER–mitochondrial interface. Currently, it is unclear whether the mitochondrial and ER motility inhibition simply stabilizes a relatively short distance between the interfacing membranes or also involves a mechanism to align the ER and OMM Ca^{2+} channels with each other.

Intermitochondrial interactions

Discrete mitochondria may function in a co-ordinated manner using Ca^{2+} or reactive oxygen species as signalling molecules (Ichas et al 1997, Pacher & Hajnoczky

2001, Zorov et al 2000, Aon et al 2003). These interactions depend on the mitochondrial density and distribution. Thus, mitochondrial movements are predicted to have some role in the control of the regenerative processes underlying Ca^{2+} or reactive oxygen species release waves.

Fusion of mitochondria can be demonstrated with isolated organelles that lack cytoskeletal support (Meeusen et al 2004) and in cells where some cytoskeletal components have been disrupted (Mattenberger et al 2003). However, normally, most of the mitochondria are anchored to the cytoskeleton and the cytoskeleton-bound mitochondria can fuse only if they were brought into contact with each other. Since mitochondria are usually aligned with a microtubule or a microfilament, organelles moving along a common track with different velocities may undergo end-to-end interactions. Mitochondria moving along separate tracks may show multiple forms of contacts, including side-to-side interactions. Thus, mitochondrial anchorage and motion will determine the sites where the fusion proteins may merge organelles with each other. Furthermore, tension has been revealed as an important factor for dynamin-mediated fission at the neck of the endocytotic vesicle (Roux et al 2006). Similarly, tension produced by the motor system may be important for the Drp1-mediated mitochondrial fission.

Biological significance of the Ca^{2+}-dependent control of the mitochondrial motility

Recent studies have revealed intracellular spatiotemporal heterogeneity in the inhibition of movement corresponding to the local $[Ca^{2+}]_c$ concentration (Yi et al 2004, Quintana et al 2006). Such differences in the movement may evoke redistribution of mitochondria to the regions of high $[Ca^{2+}]_c$. Stabilization of the position of the mitochondria in the areas of Ca^{2+} release seems to be sufficient to facilitate the delivery of $[Ca^{2+}]_c$ spikes to the mitochondria (Yi et al 2004). Targeting of mitochondria to the sites of Ca^{2+} release by Ca^{2+}-induced motility inhibition may provide the basis for a homeostatic mechanism in Ca^{2+} signalling. Retention of mitochondria at the sites of Ca^{2+} release results in an increase in the local Ca^{2+} buffering capacity. Also, mitochondrial Ca^{2+} uptake serves as a means for stimulation of mitochondrial ATP production that provides a localized energy source for Ca^{2+} reuptake by the ER/SR (Jouaville et al 1995, Landolfi et al 1998). The increase in local Ca^{2+} buffering and in ATP production represents a feedback mechanism that contributes to the control of the $[Ca^{2+}]_c$ rise. Strengthening of the Ca^{2+} scavenger mechanisms may also be important to avoid Ca^{2+}-dependent cell injury. Once Ca^{2+} release stops, motility is recovered and the mitochondria may be recruited by Ca^{2+} release in other regions of the cell. Thus, the Ca^{2+}-dependent control of mitochondrial motility offers a means to adjust the subcellular spatial organization of the Ca^{2+} buffering and energy production as needed. Ca^{2+} in combination with some other factors may underlie the synaptic activity dependent control of mitochondrial distribution in dendrites of cultured hippocampal neurons (Li et al 2004). In the same model, changes in the amount of dendritic mitochondria led to corresponding changes in the number and plasticity of spines and synapses, indicating a mutual dependence between mitochondrial dynamics and cellular plasticity.

Individual mitochondria have a limited pool of constituents and may complement each other during fusion processes. Immobilization for extended time may limit the interaction with mitochondria located in other regions of the cells. This may affect their functional stability. Furthermore, enhanced linkage of mitochondria to the ER has been shown to expose mitochondria to Ca^{2+} continuously, leading to sensitization of mitochondria to Ca^{2+}-induced membrane permeabilization and ensuing cell death (Csordas et al 2006). Thus, transient localization of mitochondria at particular regions may support cell survival but permanent immobilization may negatively affect mitochondrial function and cell survival at least in certain cell types. An interesting exception is offered by neurons, where discrete mitochondria are stationed along the axons and dendrites.

Future perspectives

This review has summarized the data indicating that the mitochondrial transport machinery senses and responds to the physiological fluctuations in $[Ca^{2+}]_c$, leading to changes in mitochondrial motility and distribution. These changes are relevant for the Ca^{2+} signal generation and for the control of calcium-dependent cell functions. Thus, the recent work on Ca^{2+} and on some other messengers has revealed a mutual interaction between cell signalling and mitochondrial motility. Identification of the molecular mechanisms of the control of mitochondrial motility remains a challenge for future work. Since some proteins involved in mitochondrial anchorage to the cytoskeletal tracks and the motors seem to be organelle specific, selective targeting of mitochondrial motions by drugs may become feasible. During the course of mitochondrial movements, local interactions are commonly formed between organelles that are in motion, involving separate tracks and motors. An intriguing task remains to sort out the mechanisms of the co-ordination of the motility of discrete mitochondria with the motility of ER domains, or with the motility of other mitochondria in the intracellular traffic control.

Acknowledgements

This work was supported by NIH grants DK51526 & GM59419 (GH). We apologize to those of our colleagues whose work could not be cited because of space restrictions.

References

Aon MA, Cortassa S, Marban E, O'Rourke B 2003 Synchronized whole cell oscillations in mitochondrial metabolism triggered by a local release of reactive oxygen species in cardiac myocytes. J Biol Chem 278:44735–44744
Beltran-Parrazal L, Lopez-Valdez H, Brennan KC et al 2006 Mitochondrial transport in processes of cortical neurons is independent of intracellular calcium. Am J Physiol Cell Physiol 291:C1193–1197
Brickley K, Smith MJ, Beck M, Stephenson FA 2005 GRIF-1 and OIP106, members of a novel gene family of coiled-coil domain proteins: association in vivo and in vitro with kinesin. J Biol Chem 280:14723–14732
Brough D, Schell MJ, Irvine RF 2005 Agonist-induced regulation of mitochondrial and endoplasmic reticulum motility. Biochem J 392:291–297
Cao TT, Chang W, Masters SE, Mooseker MS 2004 Myosin-Va binds to and mechanochemically couples microtubules to actin filaments. Mol Biol Cell 15:151–161
Carre M, Andre N, Carles G et al 2002 Tubulin is an inherent component of mitochondrial membranes that interacts with the voltage-dependent anion channel. J Biol Chem 277:33664–33669
Chada SR, Hollenbeck PJ 2004 Nerve growth factor signaling regulates motility and docking of axonal mitochondria. Curr Biol 14:1272–1276
Collins TJ, Bootman MD 2003 Mitochondria are morphologically heterogeneous within cells. J Exp Biol 206:1993–2000
Csordas G, Renken C, Varnai P et al 2006 Structural and functional features and significance of the physical linkage between ER and mitochondria. J Cell Biol 174:915–921
Deacon SW, Serpinskaya AS, Vaughan PS et al 2003 Dynactin is required for bidirectional organelle transport. J Cell Biol 160:297–301
De Vos K, Severin F, Van Herreweghe F et al 2000 Tumor necrosis factor induces hyperphosphorylation of kinesin light chain and inhibits kinesin-mediated transport of mitochondria. J Cell Biol 149:1207–1214
Fransson A, Ruusala A, Aspenstrom P 2003 Atypical Rho GTPases have roles in mitochondrial homeostasis and apoptosis. J Biol Chem 278:6495–6502
Fransson S, Ruusala A, Aspenstrom P 2006 The atypical Rho GTPases Miro-1 and Miro-2 have essential roles in mitochondrial trafficking. Biochem Biophys Res Commun 344:500–510
Frederick RL, McCaffery JM, Cunningham KW, Okamoto K, Shaw JM 2004 Yeast Miro GTPase, Gem1p, regulates mitochondrial morphology via a novel pathway. J Cell Biol 167:87–98
Glater EE, Megeath LJ, Stowers RS, Schwarz TL 2006 Axonal transport of mitochondria requires milton to recruit kinesin heavy chain and is light chain independent. J Cell Biol 173:545–557
Habermann A, Schroer TA, Griffiths G, Burkhardt JK 2001 Immunolocalization of cytoplasmic dynein and dynactin subunits in cultured macrophages: enrichment on early endocytic organelles. J Cell Sci 114:229–240
Hollenbeck PJ, Saxton WM 2005 The axonal transport of mitochondria. J Cell Sci 118:5411–5419
Hoth M, Button DC, Lewis RS 2000 Mitochondrial control of calcium-channel gating: a mechanism for sustained signaling and transcriptional activation in T lymphocytes. Proc Natl Acad Sci USA 97:10607–10612
Ichas F, Jouaville LS, Mazat JP 1997 Mitochondria are excitable organelles capable of generating and conveying electrical and calcium signals. Cell 89:1145–1153
Jouaville LS, Ichas F, Holmuhamedov EL, Camacho P, Lechleiter JD 1995 Synchronization of calcium waves by mitochondrial substrates in Xenopus laevis oocytes. Nature 377:438–441

Krementsov DN, Krementsova EB, Trybus KM 2004 Myosin V: regulation by calcium, calmodulin, and the tail domain. J Cell Biol 164:877–886

Kural C, Kim H, Syed S et al 2005 Kinesin and dynein move a peroxisome in vivo: a tug-of-war or coordinated movement? Science 308:1469–1472

Kusano H, Shimizu S, Koya RC et al 2000 Human gelsolin prevents apoptosis by inhibiting apoptotic mitochondrial changes via closing VDAC. Oncogene 19:4807–4814

Lalli G, Gschmeissner S, Schiavo G 2003 Myosin Va and microtubule-based motors are required for fast axonal retrograde transport of tetanus toxin in motor neurons. J Cell Sci 116:4639–4650

Landolfi B, Curci S, Debellis L, Pozzan T, Hofer AM 1998 Ca2+ homeostasis in the agonist-sensitive internal store: functional interactions between mitochondria and the ER measured in situ in intact cells. J Cell Biol 142:1235–1243

Li Z, Okamoto K, Hayashi Y, Sheng M 2004 The importance of dendritic mitochondria in the morphogenesis and plasticity of spines and synapses. Cell 119:873–887

Linden M, Karlsson G 1996 Identification of porin as a binding site for MAP2. Biochem Biophys Res Commun 218:833–836

Malaiyandi LM, Vergun O, Dineley KE, Reynolds IJ 2005 Direct visualization of mitochondrial zinc accumulation reveals uniporter-dependent and -independent transport mechanisms. J Neurochem 93:1242–1250

Mannella C, Renken C, Hsieh C, Marko M 2003 Cryo-electron microscopy provides evidence for tethering of mitochondria to endoplasmic reticulum. Biophys J 84:388a

Mattenberger Y, James DI, Martinou JC 2003 Fusion of mitochondria in mammalian cells is dependent on the mitochondrial inner membrane potential and independent of microtubules or actin. FEBS Lett 538:53–59

Meeusen S, McCaffery JM, Nunnari J 2004 Mitochondrial fusion intermediates revealed in vitro. Science 305:1747–1752

Milner DJ, Mavroidis M, Weisleder N, Capetanaki Y 2000 Desmin cytoskeleton linked to muscle mitochondrial distribution and respiratory function. J Cell Biol 150:1283–1298

Minin AA, Kulik AV, Gyoeva FK et al 2006 Regulation of mitochondria distribution by RhoA and formins. J Cell Sci 119:659–670

Mironov SL 2006 Spontaneous and evoked neuronal activities regulate movements of single neuronal mitochondria. Synapse 59:403–411

Morfini G, Pigino G, Beffert U, Busciglio J, Brady ST 2002 Fast axonal transport misregulation and Alzheimer's disease. Neuromolecular Med 2:89–99

Nascimento AA, Amaral RG, Bizario JC, Larson RE, Espreafico EM 1997 Subcellular localization of myosin-V in the B16 melanoma cells, a wild-type cell line for the dilute gene. Mol Biol Cell 8:1971–1988

Pacher P, Hajnoczky G 2001 Propagation of the apoptotic signal by mitochondrial waves. EMBO J 20:4107–4121

Parekh AB, Putney JW Jr 2005 Store-operated calcium channels. Physiol Rev 85:757–810

Park MK, Ashby MC, Erdemli G, Petersen OH, Tepikin AV 2001 Perinuclear, perigranular and sub-plasmalemmal mitochondria have distinct functions in the regulation of cellular calcium transport. EMBO J 20:1863–1874

Quintana A, Schwarz EC, Schwindling C, Lipp P, Kaestner L, Hoth M 2006 Sustained activity of calcium release-activated calcium channels requires translocation of mitochondria to the plasma membrane. J Biol Chem 281:40302–40309

Rice SE, Gelfand VI 2006 Paradigm lost: milton connects kinesin heavy chain to miro on mitochondria. J Cell Biol 173:459–461

Rintoul GL, Filiano AJ, Brocard JB, Kress GJ, Reynolds IJ 2003 Glutamate decreases mitochondrial size and movement in primary forebrain neurons. J Neurosci 23:7881–7888

Rizzuto R, Pinton P, Carrington W et al 1998 Close contacts with the endoplasmic reticulum as determinants of mitochondrial Ca2+ responses. Science 280:1763–1766
Roux A, Uyhazi K, Frost A, De Camilli P 2006 GTP-dependent twisting of dynamin implicates constriction and tension in membrane fission. Nature 441:528–531
Rudolf R, Mongillo M, Magalhaes PJ, Pozzan T 2004 In vivo monitoring of Ca(2+) uptake into mitochondria of mouse skeletal muscle during contraction. J Cell Biol 166:527–536
Shore GC, Tata JR 1977 Two fractions of rough endoplasmic reticulum from rat liver. I. Recovery of rapidly sedimenting endoplasmic reticulum in association with mitochondria. J Cell Biol 72:714–725
Stafford P, Brown J, Langford GM 2000 Interaction of actin- and microtubule-based motors in squid axoplasm probed with antibodies to myosin V and kinesin. Biol Bull 199:203–205
Stowers RS, Megeath LJ, Gorska-Andrzejak J, Meinertzhagen IA, Schwarz TL 2002 Axonal transport of mitochondria to synapses depends on milton, a novel Drosophila protein. Neuron 36:1063–1077
Tauhata SB, dos Santos DV, Taylor EW, Mooseker MS, Larson RE 2001 High affinity binding of brain myosin-Va to F-actin induced by calcium in the presence of ATP. J Biol Chem 276:39812–39818
Tanaka Y, Kanai Y, Okada Y et al 1998 Targeted disruption of mouse conventional kinesin heavy chain, kif5B, results in abnormal perinuclear clustering of mitochondria. Cell 93:1147–1158
Vale RD 2003 The molecular motor toolbox for intracellular transport. Cell 112:467–480
Varadi A, Johnson-Cadwell LI, Cirulli V, Yoon Y, Allan VJ, Rutter GA 2004 Cytoplasmic dynein regulates the subcellular distribution of mitochondria by controlling the recruitment of the fission factor dynamin-related protein-1. J Cell Sci 117:4389–4400
Varadi A, Tsuboi T, Rutter GA 2005 Myosin Va transports dense core secretory vesicles in pancreatic MIN6 beta-cells. Mol Biol Cell 16:2670–2680
Xu X, Forbes JG, Colombini M 2001 Actin modulates the gating of Neurospora crassa VDAC. J Membr Biol 180:73–81
Yamamoto H, Fukunaga K, Goto S, Tanaka E, Miyamoto E 1985 Ca2+, calmodulin-dependent regulation of microtubule formation via phosphorylation of microtubule-associated protein 2, tau factor, and tubulin, and comparison with the cyclic AMP-dependent phosphorylation. J Neurochem 44:759–668
Yi M, Weaver D, Hajnoczky G 2004 Control of mitochondrial motility and distribution by the calcium signal: a homeostatic circuit. J Cell Biol 167:661–672
Zorov DB, Filburn CR, Klotz LO, Zweier JL, Sollott SJ 2000 Reactive oxygen species (ROS)-induced ROS release: a new phenomenon accompanying induction of the mitochondrial permeability transition in cardiac myocytes. J Exp Med 192:1001–1014

DISCUSSION

Reynolds: We are probably substantially in agreement about most of the things we see in our cells and what you see in yours. The first studies we did involved adding huge amounts of glutamate onto cells. Lots of things happened to mitochondrial mobility and shape under these circumstances. This is a huge Ca^{2+} burden, and is a toxic situation rather than physiology. Some of the more recent studies we have done have involved more subtle signals. The basic message is that with greater activity in neurons there is less mitochondrial mobility. This is consistent with Ca^{2+} signals around the synapse slowing mitochondria down. The question I have relates to the bioenergetic state of the mitochondria in relation to the Ca^{2+} signal. In neurons we think several things happen to change mitochondrial motility. Clearly Ca^{2+} can change things around. Change in the ability of the mitochondria to produce ATP changes the motility. Then there are some intracellular signalling events that we can activate with zinc to change motility. These are quite separate. If we change the mitochondrial membrane potential, or change the activity of the ATP synthase, it stops the mitochondrial movements. We did this with NO and oligomycin, and FCCP. In your experiments in which you get Ca^{2+} signals and the mitochondria are stopping, is the mitochondrial membrane potential changing? Is this why they stop, because they are not making ATP? A follow-up question would be that you made this statement about FCCP not changing the ability of Ca^{2+} to stop the mitochondria. I am surprised by this, because I would have expected the FCCP to stop the mitochondria by itself, so that you wouldn't be able to observe the Ca^{2+} effect.

Hajnóczky: The vasopressin-induced Ca^{2+} oscillations and the ensuing mitochondrial motility inhibition were not associated with an apparent change in the mitochondrial membrane potential. FCCP or other uncouplers caused a rapid loss in the membrane potential and a progressive decay in motility. When FCCP was combined with oligomycin to prevent the mitochondrial ATP consumption, the membrane potential dissipated but the decay in the motility was relatively slow. During the period when the mitochondrial membrane potential was gone but the motility was maintained vasopressin could stop the mitochondria. Thus, the vasopressin-induced decrease in mitochondrial motility was not dependent on the mitochondrial membrane potential.

Jacobs: I was going to ask about the local concentrations of ADP and ATP in the vicinity of where mitochondria have stopped moving because of a local Ca^{2+} signal. Do I understand from what you have just said that you have evidence that ADP and ATP are not involved in restraining the movement of mitochondria under the Ca^{2+} signalling? I'm not clear why you are stating that ADP is not the mechanism. In other words, that there is a direct effect of Ca^{2+} on some sensor, rather than an ADP sensor.

Hajnóczky: The Ca^{2+} signal is likely to be associated with some changes in ATP and ADP levels, which have been claimed to affect the mitochondrial motility (Brough et al 2005, Mironov 2006). However, the Ca^{2+} effect is not dependent on a change in ADP or ATP concentration. The strongest evidence for this point is that in permeabilized cells, we could add an ATP-regenerating system and still detect Ca^{2+}-induced mitochondrial movement inhibition. I would predict a direct effect of Ca^{2+} on some sensor. As I mentioned in the talk, miro and myosin Va are candidates for the Ca^{2+} sensor.

Rizzuto: We have some data suggesting that this process is probably operated by the mitochondrial GTPase miro. We have data in neurons and non-excitable cells using silenced miro, and with a miro mutant that doesn't bind Ca^{2+}, indicating that miro does the job. There is a direct effect of Ca^{2+}, and we have sufficient evidence to say that it is miro.

Scorrano: Do you envisage a situation in which microfilaments can act as a stop signal for mitochondria? You are describing what moves the mitochondria, but somehow mitochondria should be stopped where the cell would like them to be, if we reason teleologically.

Hajnóczky: For mitochondria moving along microtubules, cross-linkage to the microfilaments is a distinct possibility for the motility inhibition. Myosin Va both binds to the mitochondria and displays a Ca^{2+}-dependent interaction with actin-filaments (Tauhata et al 2001, Krementsov et al 2004). Regarding the second question, stopping mitochondria at the sites where [Ca^{2+}] rises or [ATP] decreases may provide a mechanism to handle the Ca^{2+} load and regenerate the ATP.

Scorrano: What about intermediate filaments?

Hajnóczky: An interesting observation has been made on the interaction of mitochondria and desmin, and on the significance of this interaction in mitochondrial function (Milner et al 2000).

Adam-Vizi: You showed that when mitochondria stop moving they take up more Ca^{2+}. You used nocodazole and applied a Ca^{2+} signal, and there was higher Ca^{2+} uptake by mitochondria. But at the same time after nocodazole treatment mitochondria seemed to be more depolarized. How do you reconcile these observations?

Duchen: The oligomycin experiment with nocodazole suggests that potential is being maintained by reversal of the ATPase in these neurons. Doesn't this suggest that their respiration is being compromised?

Hajnóczky: Nocodazole was used to evaluate the effect of both the short and long term motility inhibition on mitochondrial function. When the effect of mitochondrial motility on the IP_3 receptor-mitochondrial Ca^{2+} transfer was evaluated, nocodazole was added for 10–20 min. No mitochondrial depolarization was observed in the nocodazole-treated cells for 20 min. Also, the cells were not sensitized to oligomycin-induced depolarization in the course of these experiments. Therefore nocodazole did not seem to alter the driving force for the mitochondrial Ca^{2+} uptake.

Even 3–4 h pretreatment with nocodazole did not suppress the uptake of the potentiometric probes by the mitochondria. However, 3–4 h nocodazole pretreatment sensitized the mitochondria to the oligomycin-induced depolarization, indicating that a large fraction of mitochondria became ATP consumers. We propose that this change in mitochondrial metabolism is not necessarily elicited by a toxic effect of nocodazole, but may be a gradually developing consequence of the inhibition of the mitochondrial fusion-fission dynamics that depends on the microtubules.

Duchen: You suggested that the ER and mitochondria can move together, at least in some parts of the cell. How does this work? Is there any mechanism for ER to move with motors?

Hajnóczky: First, both ER and mitochondria may have their own motors that are synchronized with each other. Alternatively, the two organelles are bound together allowing a 'piggy back' mechanism. In a recent study, we have shown direct physical links, termed as 'tethers' between ER and mitochondria (Csordás et al 2006). If the attachment points have lateral mobility in the ER membrane, the tethers may support the movement of mitochondria along the ER where the ER is continuous, and may enable ER and mitochondria to move together where discrete pieces of ER are attached to the moving mitochondria.

Halestrap: If I understood you correctly, if you give low vasopressin, you have transient Ca^{2+} spikes. Are you saying that over the first few transients the mitochondria would recruit to the ER progressively as they lock in at each Ca^{2+} spike? And if this is the case, would you not see a change in the mitochondrial Ca^{2+} signal in the first few vasopressin spikes? Is that observed?

Hajnóczky: I would predict that spatially confined Ca^{2+} transients may cause recruitment of the mitochondria to the sites of the Ca^{2+} release. I am not sure that a few single spikes would be enough to evoke a striking change in the mitochondrial distribution. However, in paradigms of store-operated Ca^{2+}, entry redistribution of mitochondria to the plasma membrane has been nicely demonstrated (Malli et al 2005, Quintana et al 2006).

The time course of the mitochondrial Ca^{2+} signal seems to indicate the presence of a gradual improvement of the IP_3 receptor-mitochondrial Ca^{2+} transfer in several cell types. However, this facilitation step may also take place at the level of the uniporter's Ca^{2+} permeability (Csordás & Hajnóczky 2003).

Bernardi: In your model, both on microtubules and ER, you are thinking of the mitochondria rolling over the surface on a track. What is providing a stable enough attachment site to prevent unattaching, yet still allow movement?

Hajnóczky: As to the connection to the microtubules, the miro contains a transmembrane domain at the C-terminus, which is essential for the mitochondrial targeting of the protein. The protein(s) forming the ER–mitochondrial tethers remains to be identified.

Brand: Your cells are presumably under oxidative stress because they are exposed to atmospheric oxygen. If one function of this system is kiss-and-run to separate off oxidative damage, do you think you are seeing a situation that is violently trying to respond to this horrible environment, and which wouldn't normally be doing this stuff to the same extent *in vivo*?

Hajnóczky: That's a difficult question. The cell cultures were also grown in the presence of atmospheric oxygen.

Reynolds: The cells that are living through the atmospheric oxygen are the ones that have been selected for their ability to survive in that environment.

References

Brough D, Schell MJ, Irvine RF 2005 Agonist-induced regulation of mitochondrial and endoplasmic reticulum motility. Biochem J 392:291–297
Csordás G, Hajnóczky G 2003 Plasticity of mitochondrial calcium signaling. J Biol Chem 278:42273–42782
Csordás G, Renken C, Várnai P et al 2006 Structural and functional features and significance of the physical linkage between ER and mitochondria. J Cell Biol 174:915–921
Krementsov DN, Krementsova EB, Trybus KM 2004 Myosin V: regulation by calcium, calmodulin, and the tail domain. J Cell Biol 164:877–886
Malli R, Frieden M, Trenker M, Graier WF 2005 The role of mitochondria for Ca^{2+} refilling of the endoplasmic reticulum. J Biol Chem 280:12114–12122
Milner DJ, Mavroidis M, Weisleder N, Capetanaki Y 2000 Desmin cytoskeleton linked to muscle mitochondrial distribution and respiratory function. J Cell Biol 150:1283–1298
Mironov SL 2006 Spontaneous and evoked neuronal activities regulate movements of single neuronal mitochondria. Synapse 59:403–411
Quintana A, Schwarz EC, Schwindling C, Lipp P, Kaestner L, Hoth M 2006 Sustained activity of calcium release-activated calcium channels requires translocation of mitochondria to the plasma membrane. J Biol Chem 281:40302–40309
Tauhata SB, dos Santos DV, Taylor EW, Mooseker MS, Larson RE 2001 High affinity binding of brain myosin-Va to F-actin induced by calcium in the presence of ATP. J Biol Chem 276:39812–39818

Endoplasmic reticulum/mitochondria calcium cross-talk

Anna Romagnoli, Paola Aguiari, Diego De Stefani, Sara Leo, Saverio Marchi, Alessandro Rimessi, Erika Zecchini, Paolo Pinton and Rosario Rizzuto

ER-GenTech and Interdisciplinary Center for the Study of Inflammation, Department of Experimental and Diagnostic Medicine, University of Ferrara, Italy

Abstract. The interaction of mitochondria with the endoplasmic reticulum (ER) Ca^{2+} store plays a key role in allowing these organelles to rapidly and effectively respond to cellular Ca^{2+} signals. In this contribution, we will briefly discuss: (i) old and new concepts of mitochondrial Ca^{2+} homeostasis; (ii) the relationship between mitochondrial 3D structure and Ca^{2+} homeostasis; (iii) the modulation by cytoplasmic signalling pathways; and (iv) new data suggesting that mitochondria and ER Ca^{2+} channels are assembled in a macromolecular complex in which the inositol-1,4,5-trisphosphate receptor directly stimulates the mitochondrial Ca^{2+} uptake machinery.

2007 Mitochondrial biology: new perspectives. Wiley, Chichester (Novartis Foundation Symposium 287) p 122–139

Mitochondrial Ca^{2+} uptake: an old history of a new concept

A long-standing biological observation is the capacity of energized mitochondria to transport Ca^{2+} across the ion-impermeable inner membrane. Indeed, the translocation of protons by the respiratory chain complexes establishes a large electrochemical gradient, mostly composed of an electrical gradient ($\Delta\Psi$), that represents a large thermodynamic force for Ca^{2+} accumulation into the mitochondrial matrix. Work carried out in isolated mitochondria in the 1960s and 70s characterized the fundamental properties of Ca^{2+} transport in mitochondria, which, apart from the recent electrophysiological demonstration that the mitochondrial Ca^{2+} uniporter (MCU) is a bona fide channel (Kirichok et al 2004), still represents the current knowledge of the process. Ca^{2+} is accumulated into the matrix through a Ruthenium red-sensitive electrogenic route (the MCU) and re-extruded, in exchange with monovalent cations (H^+ or Na^+), by two antiporters that prevent the attainment of electrical equilibrium (that would imply, for a mitochondrial membrane potential, $\Delta\Psi_m$, of 180 mV and a cytosolic Ca^{2+} concentration of 0.1 µM, accumulation of Ca^{2+} into the matrix up to 0.1 M). It was thus logical to assume that mitochondria were loaded with Ca^{2+}, possibly releasing it in a number of physiological and/or pathological conditions.

However, when in the 1980s Ca^{2+} homeostasis emerged as a ubiquitous signalling route, characterized by a remarkable spatiotemporal and molecular complexity, the role of mitochondria took the opposite route, and was greatly downplayed to that of a low-affinity, high capacity sink coming into action only in the case of major cellular overload with Ca^{2+} (i.e. in pathophysiological conditions). Indeed, it became apparent that the endoplasmic reticulum (ER), and not the mitochondria, is the source of rapidly mobilizable Ca^{2+}: it possesses a Ca^{2+} pump for accumulating Ca^{2+}, and the inositol-1,4,5-trisphosphate receptor (IP_3R) for rapidly releasing it upon cell stimulation. At the same time, mitochondria did not appear an important target of the released Ca^{2+}, as the $[Ca^{2+}]$ reached in the cytoplasm (2–3 µM at the peak) appeared well below that required for rapid accumulation by the low-affinity MCU.

This situation was completely reversed when novel gene-encoded targeted probes allowed the unambiguous measurement of the $[Ca^{2+}]$ of the mitochondrial matrix ($[Ca^{2+}]_m$). This was first achieved by targeting to mitochondria a Ca^{2+}-sensitive photoprotein, aequorin, which demonstrated that a rapid $[Ca^{2+}]_m$ peak, reaching values well above those of the bulk cytosol, parallels the $[Ca^{2+}]$ rise evoked in the cytoplasm by cell stimulation (Rizzuto et al 1992). Similar conclusions could be reached also with fluorescent indicators, such as the positively charged Ca^{2+} indicator rhod-2 (that accumulates within the organelle) (Jou et al 1996) and the more recently developed GFP-based fluorescent indicators (Nagai et al 2001). With the latter probes, endowed with a much stronger signal than the photoprotein, single cell imaging of organelle Ca^{2+} can be carried out. Thus, it is possible to match the accurate estimates of $[Ca^{2+}]_m$ values, obtained with the photoprotein, with detailed spatiotemporal analyses of $[Ca^{2+}]_m$ transients. With these tools in hands, it was confirmed that mitochondria promptly respond to cytosolic $[Ca^{2+}]$ rises, and also that the $[Ca^{2+}]_c$ oscillations, the typical response to agonists of many cell types, are paralleled by rapid spiking of $[Ca^{2+}]_m$, thus specifically decoding a frequency-mediated signal within the mitochondria, as clearly shown in hepatocytes (Thomas et al 1995), cardiomyocytes (Trollinger et al 1997) and HeLa cells (Rizzuto et al 1994).

The rediscovery of the process of mitochondrial Ca^{2+} uptake has been paralleled by the appreciation of its role in regulating widely diverse cellular functions. Within the matrix two radically different effects can be triggered by a $[Ca^{2+}]$ rise. The first, as originally proposed by Denton, McCormack and Hansford in the 1960s, is the activation of three key metabolic enzymes (the pyruvate, α-ketoglutarate and isocitrate dehydrogenases), thus stimulating aerobic metabolism when a cell is stimulated to perform energy-consuming processes in the cytosol (e.g. contraction, secretion, etc.). Indeed, the direct measurement of mitochondrial

ATP levels with a targeted chimera of the ATP-sensitive photoprotein luciferase demonstrated that the Ca^{2+} signal within the mitochondria is responsible for the enhanced ATP production, an effect that lasts longer than the Ca^{2+} signal itself, highlighting a novel form of cellular 'metabolic memory' (Jouaville et al 1999). In some conditions, however, a Ca^{2+} signal within the mitochondria may trigger cell death. The alteration of the Ca^{2+} signal reaching the mitochondria and/or the combined action of apoptotic agents or pathophysiological conditions (e.g. oxidative stress) can induce a profound alteration of organelle structure and function (Szalai et al 1999, Pinton et al 2001). As a consequence, proteins normally retained in the organelle, such as an important component of the respiratory chain, cytochrome c, (Kluck et al 1997, Yang et al 1997), as well as more recently discovered proteins, such as AIF (Susin et al 1999) and Smac/Diablo (Du et al 2000), are released into the cytoplasm, where they activate effector caspases and drive cells to apoptotic cell death (Fig. 1). In relation to this effect, the anti-oncogene Bcl-2 was shown to reduce the steady state Ca^{2+} levels in the ER (and thus dampen the pro-apoptotic Ca^{2+} signal) (Pinton et al 2000, Foyouzi-Youssefi et al 2000).

On the cytosolic side, mitochondrial Ca^{2+} uptake exerts two different effects. In the first, the spatial clustering of mitochondria in a defined portion of the cell represents a physiological 'fixed spatial buffer' that prevents (or delays) the spread of cytoplasmic Ca^{2+} waves, as elegantly demonstrated in pancreatic acinar cells. In these cells, the Ca^{2+} response to a low-dose agonist stimulation is restricted to the apical pole (where it causes granule secretion) by the action of a mitochondrial 'firewall' located between the apical and basolateral portions of the cell (Tinel et al 1999). Mitochondria, however, can affect cytosolic Ca^{2+} signalling through a different mechanism. Indeed, since they are located in close proximity to ER (or plasma membrane) channels that avidly take up part of the Ca^{2+} released upon stimulation, they reduce the $[Ca^{2+}]$ in the critical microenvironment of the open Ca^{2+} channel. In this way, they modulate the feedback effect of Ca^{2+} (negative or positive, depending on the Ca^{2+} concentrations and channel isoform) on the channel itself, and thus the duration and amplitude of ER Ca^{2+} release. This mechanism was first demonstrated in *Xenopus* oocytes, in which the energization state (and thus the capacity to accumulate Ca^{2+}) was shown to influence the spatiotemporal pattern of the typical propagating Ca^{2+} waves induced by IP_3 (Jouaville et al 1995). Then, a series of studies in mammalian cells confirmed the notion and demonstrated its importance in shaping the Ca^{2+} responses to agonists of cells as diverse as hepatocytes (Thomas et al 1995) astrocytes (Boitier et al 1999) or BHK cells (Landolfi et al 1998). In addition, the concept has been extended to the control of plasma membrane channels, as in the case of the relief of the Ca^{2+}-dependent inhibition of CRAC channels, i.e. the influx pathway triggered by the drop of $[Ca^{2+}]$ in intracellular stores (Hoth et al 1997, Gilabert & Parekh 2000).

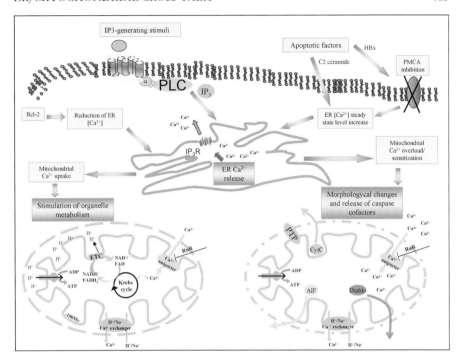

FIG. 1. Mitochondrial [Ca^{2+}] rise can trigger two radically different effects: stimulation of aerobic metabolism or cell death. IP_3-generating stimuli at cell membrane provide IP_3R opening, causing Ca^{2+} release from ER. Mitochondria closely located near ER channels take up part of the Ca^{2+} released through the mitochondrial Ca^{2+} uniporter (MCU). Ca^{2+} in the matrix activates key metabolic enzymes in the Krebs cycle thus stimulating organelle metabolism. Apoptotic factors exert their effects increasing the amount of Ca^{2+} released from ER, directly (such as the lipid mediator of apoptosis C2-ceramide) or indirectly, i.e. the Herpetic B virus X protein HBx which inhibits PMCA activity. Mitochondrial Ca^{2+} overload determines PTP opening and release of mitochondrial proteins such as cytochrome c, AIF or Smac/Diablo in the cytosol, triggering activation of apoptotic pathways. Conversely, Bcl-2 exerts its anti-apoptotic effect reducing ER steady state level, thus preventing mitochondrial Ca^{2+} overload.

The regulation of mitochondrial Ca^{2+} homeostasis

Given the growing interest in the process of mitochondrial Ca^{2+} homeostasis, the focus of this contribution is the clarification of the mechanism that finely tunes the Ca^{2+} transfer from the ER to the mitochondria. Indeed, a dynamic membrane-bound organelle with a compound protein machinery for Ca^{2+} uptake and release implies that various mechanisms could co-operate in regulating the Ca^{2+} responsiveness of mitochondria (Fig. 2). We will discuss our own data, pointing to three different mechanisms:

— the three dimensional structure of the organelle
— signalling pathways activated by extracellular stimuli
— protein–protein interactions within macromolecular complexes assembled at ER–mitochondria contacts.

The 3D structure of mitochondria

The observation that mitochondria in different cell types, or within the same cell in different physiological or pathological conditions, can form a largely connected network extended throughout the cell or represent an isolated oval-shaped organelle is as old as the microscopic identification of this organelle. However, only in recent years has the control of mitochondrial shape emerged as a tightly regulated cellular process, with a dedicated molecular machinery. Indeed, not only have the proteins promoting mitochondrial fusion or fission been identified and shown to be membrane-associated GTP-binding mechanochemical enzymes, but also these proteins and their activities have been shown to play an important role in determining the sensitivity of cells to apoptotic challenges. While this excites further interest in the activity of these proteins, the mechanism through which alterations of mitochondrial shape can affect mitochondrial participation in apoptosis is still debated. The modulation of Ca^{2+} signals represented an obvious possibility, so we investigated Ca^{2+} and apoptotic changes in HeLa cells in which the 3D structure of mitochondria was modified through recombinant expression of the mitochondrial fission factor dynamin-related protein 1 (Drp1) (Frank et al 2001). Ca^{2+} homeostasis in the cytosol and in mitochondria was investigated by two approaches, i.e. the co-transfection of aequorin and the measurement of $[Ca^{2+}]$ in the whole cell population (thus obtaining an accurate estimate of the global phenomenon) and the co-transfection of GFP-based recombinant Ca^{2+} probes that allow us to carry out single cell imaging experiments with high spatial and temporal resolution. The imaging results showed that waves of $[Ca^{2+}]_m$ originate from distinct sites (most likely corresponding to the contacts with the ER) and travel across the network. In fragmented mitochondria, wave diffusion is blocked, mitochondrial Ca^{2+} increases are smaller, and thus the sensitivity to Ca^{2+}-dependent apoptotic challenges (such as the lipid mediator of apoptosis C2-ceramide) is reduced (Szabadkai et al 2004).

Modulation by cytosolic signalling pathways

On this topic, we will describe only one interesting example, but we remind the reader that several other pathways besides that described here are emerging that affect mitochondrial Ca^{2+} responsiveness (see for example the different effects of the various PKC isoforms on cytosolic and mitochondrial Ca^{2+} homeostasis;

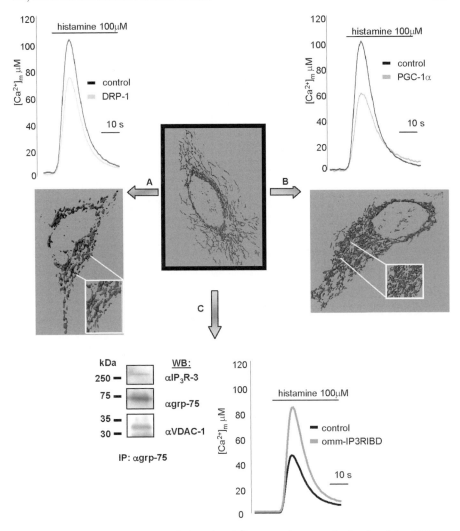

FIG. 2. Three mechanisms regulating the Ca^{2+} responsiveness of mitochondria: different physio/pathological conditions can directly modulate mitochondrial Ca^{2+} uptake with different mechanisms. (A) Mitochondrial network fragmentation induced by the fission factor DRP1 (lower panel) reduces $[Ca^{2+}]_m$ in response to histamine stimulation (upper panel). (B) The reduction in $[Ca^{2+}]_m$ value after agonist stimulation (upper panel) is also observed in cells overexpressing PGC1α due to increasing in mitochondrial volume (lower panel). (C) Mitochondrial Ca^{2+} uptake can also be modulated by protein–protein interactions: ER and mitochondrial Ca^{2+} channels are coupled through the molecular chaperone Grp75 at ER–mitochondria contact sites (as demonstrated by immunoprecipitation on the left panel); moreover, expression of the ligand binding domain of the IP_3R1 tethered to the outer mitochondrial membrane (omm-IP3R1BD) can enhance mitochondrial Ca^{2+} uptake (see traces on the right panel). The 3D reconstructions were obtained from HeLa cells transfected with mtRFP for control (cell in the middle) and co-transfected with DRP1 or PGC1α (on the left and right, respectively). $[Ca^{2+}]_m$ traces were carried out measuring aequorin luminescence in a population of transfected HeLa cells with mtAEQmut for control and co-transfected with DRP1 or PGC1α.

Pinton et al 2004). We have investigated the effect of the complex mitochondrial changes triggered by the genomic programme activated by the transcriptional coactivator peroxisome proliferator-activated receptor γ (PPARγ) coactivator 1α (PGC1α). By integrating the action of several transcription factors (Puigserver & Spiegelman 2003) PGC1α orchestrates the response to several physiological and pathological stimuli, such as cold exposure, fasting and muscle exercise. Its main effect on mitochondria is the stimulation of mitochondrial biogenesis in concert with the increased expression of electron transport chain (ETC) components and uncoupling proteins (UCP1/2) (Nakatani et al 2002, Puigserver et al 1998). This compound effect increases both the respiratory capacity of mitochondria and the leakiness of the inner membrane (therefore reducing mitochondrial 'efficiency'), thus accounting for the thermogenic response of brown fat and skeletal muscle. Conversely, the functional relevance in non-thermogenic tissues (and thus the role in other physiological conditions) is less clear. By using aequorin and GFP-based probes, we demonstrated that the $[Ca^{2+}]_m$ increases evoked by cell stimulation are markedly reduced in PGC1α expressing cells. As the number of ER/mitochondrial contacts appears unchanged, this effect is at least in part due to mitochondrial biogenesis, and thus redistribution of the Ca^{2+} load in a larger volume (Bianchi et al 2006). However, it is likely that the proteomic changes triggered by PGC1α overexpression also directly involve the still undefined machinery for mitochondrial Ca^{2+} uptake. Based on this intriguing possibility, we are searching *in silico*, among the genes up- or down-regulated by PGC1α, for potential candidates for the role of mitochondrial Ca^{2+} transporters and/or regulators. Finally, the reduction of mitochondrial Ca^{2+} responses is paralleled by reduced sensitivity to apoptotic challenges, a result that nicely matches the observation that *PGC1α$^{-/-}$* shows alterations in cell death pathways (see the lesions in the striatal region leading to hyperactivity and thus an unexpected lean phenotype) (Lin et al 2004).

Protein–protein interactions between ER and mitochondrial Ca^{2+} channels

As already explained, our lack of molecular understanding of mitochondrial Ca^{2+} homeostasis is almost complete. Indeed, neither the MCU nor the exchangers have been identified. The only transporter with an established role is the voltage-dependent anion channel (VDAC) of the outer mitochondrial membrane (OMM) (Rapizzi et al 2002). Indeed, although it is located in the ion-permeable OMM, data obtained by our group demonstrated that it plays an important role in allowing the Ca^{2+} microdomain generated close to the IP_3R to rapidly diffuse to the MCU, and thus maximize mitochondrial Ca^{2+} uptake. Indeed, the VDAC repertoire proved to be a key determinant of mitochondrial Ca^{2+} responsiveness. Based on these results, we hypothesized that VDAC could be part of larger signalling complex with ER and/or IMM proteins. We thus initiated a two-hybrid search of

VDAC interactors, with the aim of identifying other proteins that have a role in mitochondrial Ca^{2+} homeostasis. Among the proteins that were identified, the chaperone Grp75 could be shown to interact with the IP_3R and VDAC itself, placing the two channels in close molecular proximity. Interestingly, this molecular proximity allowed the IP3BD of the IP_3R to directly stimulate mitochondrial Ca^{2+} uptake, not only when Ca^{2+} is released from the ER but also in conditions in which the Ca^{2+} rise originates from the plasma membrane. Thus, the conclusion could be drawn that the direct interaction of the Ca^{2+} channels of the two organelles, in turn controlled by a partly cysotolic chaperone, is an additional checkpoint controlling the Ca^{2+} cross-talk between the two organelles.

Conclusions

The sites of close interaction between the ER and mitochondria appear to play a role in the participation of mitochondria in Ca^{2+} signalling. On the one hand, both organelles participate in the control of the $[Ca^{2+}]$ in this restricted space: the ER releases Ca^{2+} that reaches a concentration allowing the low-affinity MCU to rapidly accumulate Ca^{2+} in the matrix, and mitochondria clear part of the Ca^{2+}, thus modulating the feedback control of the cation on the IP_3R. In addition, the IP_3R and VDAC are scaffolded in a macromolecular complex, in which their direct interaction alters the efficiency of mitochondrial Ca^{2+} uptake. Finally, a number of signalling pathways, as well as the fusion/fission state of mitochondria, also participate in regulating mitochondrial responsiveness to cytoplasmic Ca^{2+} signals. Much more work will be needed to clarify the mechanisms, as well as the relative importance, of all these effects, but when completed this analysis will allow us to understand how mitochondria can translate Ca^{2+} signals into very different cellular functions, and which molecules can be designed to correct the dysfunctions occurring in pathophysiological conditions.

Acknowledgements

Experimental work in the authors' laboratory was supported by grants from the Italian University Ministry (PRIN, FIRB), Telethon-Italy (grant #GGP05284), the Italian Association for Cancer Research (AIRC), the Italian Space Agency (ASI), the EU NeuroNE network, the PRRIITT programme of the Emilia-Romagna region and 'Obiettivo 2' funds from the Ferrara province.

References

Bianchi K, Vandecasteele G, Carli C, Romagnoli A, Szabadkai G, Rizzuto R 2006 Regulation of Ca2+ signalling and Ca2+-mediated cell death by the transcriptional coactivator PGC-1alpha. Cell Death Differ 13:586–596

Boitier E, Rea R, Duchen MR 1999 Mitochondria exert a negative feedback on the propagation of intracellular Ca2+ waves in rat cortical astrocytes. J Cell Biol 145:795–808

Du C, Fang M, Li Y, Li L, Wang X 2000 Smac, a mitochondrial protein that promotes cytochrome c-dependent caspase activation by eliminating IAP inhibition. Cell 102:33–42

Foyouzi-Youssefi R, Arnaudeau S, Borner C et al 2000 Bcl-2 decreases the free Ca2+ concentration within the endoplasmic reticulum. Proc Natl Acad Sci USA 97:5723–5728

Frank S, Gaume B, Bergmann-Leitner ES et al 2001 The role of dynamin-related protein 1, a mediator of mitochondrial fission, in apoptosis. Dev Cell 1:515–525

Gilabert JA, Parekh AB 2000 Respiring mitochondria determine the pattern of activation and inactivation of the store-operated Ca(2+) current I (CRAC). EMBO J 19:6401–6407

Hoth M, Fanger CM, Lewis RS 1997 Mitochondrial regulation of store-operated calcium signaling in T lymphocytes. J Cell Biol 137:633–648

Jou MJ, Peng TI, Sheu SS 1996 Histamine induces oscillations of mitochondrial free Ca2+ concentration in single cultured rat brain astrocytes. J Physiol 497:299–308

Jouaville LS, Ichas F, Holmuhamedov EL, Camacho P, Lechleiter JD 1995 Synchronization of calcium waves by mitochondrial substrates in Xenopus laevis oocytes. Nature 377:438–441

Jouaville LS, Pinton P, Bastianutto C, Rutter GA, Rizzuto R 1999 Regulation of mitochondrial ATP synthesis by calcium: evidence for a long-term metabolic priming. Proc Natl Acad Sci USA 96:13807–13812

Kirichok Y, Krapivinsky G, Clapham DE 2004 The mitochondrial calcium uniporter is a highly selective ion channel. Nature 427:360–364

Kluck RM, Bossy-Wetzel E, Green DR, Newmeyer DD 1997 The release of cytochrome c from mitochondria: a primary site for Bcl-2 regulation of apoptosis. Science 275:1132–1136

Landolfi B, Curci S, Debellis L, Pozzan T, Hofer AM 1998 Ca2+ homeostasis in the agonist-sensitive internal store: functional interactions between mitochondria and the ER measured in situ in intact cells. J Cell Biol 142:1235–1243

Lin J, Wu PH, Tarr PT et al 2004 Defects in adaptive energy metabolism with CNS-linked hyperactivity in PGC-1alpha null mice. Cell 119:121–135

Nagai T, Sawano A, Park ES, Miyawaki A 2001 Circularly permuted green fluorescent proteins engineered to sense Ca2+. Proc Natl Acad Sci USA 98:3197–3202

Nakatani T, Tsuboyama-Kasaoka N, Takahashi M, Miura S, Ezaki O 2002 Mechanism for peroxisome proliferator-activated receptor-alpha activator-induced up-regulation of UCP2 mRNA in rodent hepatocytes. J Biol Chem 277:9562–9569

Pinton P, Ferrari D, Magalhaes P et al 2000 Reduced loading of intracellular Ca(2+) stores and downregulation of capacitative Ca(2+) influx in Bcl-2-overexpressing cells. J Cell Biol 148:857–862

Pinton P, Ferrari D, Rapizzi E, Di Virgilio F, Pozzan T, Rizzuto R 2001 The Ca2+ concentration of the endoplasmic reticulum is a key determinant of ceramide-induced apoptosis: significance for the molecular mechanism of Bcl-2 action. EMBO J 20:2690–2701

Pinton P, Leo S, Wieckowski MR, Di Benedetto G, Rizzuto R 2004 Long-term modulation of mitochondrial Ca2+ signals by protein kinase C isozymes. J Cell Biol 165:223–232

Puigserver P, Spiegelman BM 2003 Peroxisome proliferator-activated receptor-gamma coactivator 1 alpha (PGC-1 alpha): transcriptional coactivator and metabolic regulator. Endocr Rev 24:78–90

Puigserver P, Wu Z, Park CW, Graves R, Wright M, Spiegelman BM 1998 A cold-inducible coactivator of nuclear receptors linked to adaptive thermogenesis. Cell 92:829–839

Rapizzi E, Pinton P, Szabadkai G et al 2002 Recombinant expression of the voltage-dependent anion channel enhances the transfer of Ca2+ microdomains to mitochondria. J Cell Biol 159:613–624

Rizzuto R, Simpson AW, Brini M, Pozzan T 1992 Rapid changes of mitochondrial Ca2+ revealed by specifically targeted recombinant aequorin. Nature 358:325–327

Rizzuto R, Bastianutto C, Brini M, Murgia M, Pozzan T 1994 Mitochondrial Ca2+ homeostasis in intact cells. J Cell Biol 126:1183–1194

Susin SA, Lorenzo HK, Zamzami N et al 1999 Molecular characterization of mitochondrial apoptosis-inducing factor. Nature 397:441–446

Szabadkai G, Simoni AM, Chami M, Wieckowski MR, Youle RJ, Rizzuto R 2004 Drp-1-dependent division of the mitochondrial network blocks intraorganellar Ca2+ waves and protects against Ca2+-mediated apoptosis. Mol Cell 16:59–68

Szalai G, Krishnamurthy R, Hajnoczky G 1999 Apoptosis driven by IP(3)-linked mitochondrial calcium signals. EMBO J 18:6349–6361

Thomas AP, Renard-Rooney DC, Hajnoczky G, Robb-Gaspers LD, Lin C, Rooney TA 1995 Subcellular organization of calcium signalling in hepatocytes and the intact liver. In: Calcium waves, gradients and oscillations. Wiley, Chichester (Ciba Found Symp 188) p 18–35

Tinel H, Cancela JM, Mogami H et al 1999 Active mitochondria surrounding the pancreatic acinar granule region prevent spreading of inositol trisphosphate-evoked local cytosolic Ca(2+) signals. EMBO J 18:4999–5008

Trollinger DR, Cascio WE, Lemasters JJ 1997 Selective loading of Rhod 2 into mitochondria shows mitochondrial Ca2+ transients during the contractile cycle in adult rabbit cardiac myocytes. Biochem Biophys Res Commum 236:738–742

Yang J, Liu X, Bhalla K et al 1997 Prevention of apoptosis by Bcl-2: release of cytochrome c from mitochondria blocked. Science 275:1129–1132

DISCUSSION

Nicholls: It has worried me for a long time that studies on mitochondrial Ca^{2+} transport seem to have been going along two independent but parallel lines. There is the area with non-excitable cells where we are dealing with the interaction between ER and mitochondria. But what we neuronal people find is that when we measure Ca^{2+} uptake by, for example, isolated brain mitochondria and calculate the activity of the Ca^{2+} uniporter, we do not see a 'low-affinity' transporter. The activity of the uniporter activity increases as the 2.5 power of the cytoplasmic Ca^{2+}. When the extramitochondrial Ca^{2+} rises above $0.5\,\mu M$, uptake exceeds efflux, as can be seen by repetitive Ca^{2+} addition experiments with isolated mitochondria. By the time levels of $3\,\mu M$ are reached, all the respiratory capacity is devoted to Ca^{2+} uptake.

Experiments that Michael Duchen's group and others have done with mitochondrial transport in intact neurons and neural cells shows a precise correlation between what was shown with isolated mitochondria and what occurs in intact neurons. We need to accept that cell type has an enormous influence over what is actually happening. The ER–mitochondria collaboration plays a major role in some cells, and in neuronal cells with mitochondria moving on these microtubular tracks the transport properties fit closely to those of isolated mitochondria.

Rizzuto: I agree. To account for the rapid responsiveness in non-excitable cells, with these pulsatile changes, you have to account for a microdomain. This does not rule out the fact that in neurons and neuronal dendrites secretion is exactly as you say. Also, in a non-excitable cell which undergoes a sustained rise, such as Spat's data on long-term steroid production in luteal cells, there is a small Ca^{2+} increase. We don't need high responsiveness to Ca^{2+}. Your point is well taken: we are probably dealing with the most common signalling method in non-excitable cells. But we have to account for complexity, and recognize that in other cells the cells can simply respond to the bulk cytosolic rise.

Spiegelman: To what extent is the Ca^{2+} uptake by mitochondria a kinetic effect, that is, an equilibrium or steady-state effect influenced by the membrane potential of the mitochondria? You have presented this as though it is controlled by uptake via the pores, and there is basically infinite capacity. To what extent is the uptake limited, not just by VDAC or by the uniporter, but also by the membrane potential of the mitochondrion itself?

Rizzuto: There are some nice data from Ole Petersen and his group in Liverpool. They have made a correlation between spiking and the changes in membrane potential in a pancreatic acinar cell. He sees minimal drops in the membrane potential for every Ca^{2+} spike, and this is physiological and so there is a relatively small Ca^{2+} pulse. In physiological signalling the net amount of Ca^{2+} that goes in does not change $\Delta\Psi$ as significantly as to make this a regulatory element.

Spiegelman: Are you saying that it is so negatively charged on the inside of the mitochondria it is a sink?

Lemasters: What about the converse: is a decrease in $\Delta\Psi$ going to slow Ca^{2+} up?

Rizzuto: We haven't looked in detail. We have done some work on cybrids harbouring mitochondrial mutations showing that this is the case.

Nicholls: The uptake is rather independent of membrane potential; it is not a thermodynamic equilibrium. This is another point where there is controversy. Work by Richard Hansford, Richard Denton and the Bristol group came up with the idea that free matrix Ca^{2+} concentration hovers around the $1\,\mu M$ level. This is what is needed to activate the matrix dehydrogenases. We confirmed this a couple of years ago with brain mitochondria. Thus there is no significant free Ca^{2+} concentration gradient across the inner mitochondrial membrane, although vast amounts of *total* Ca^{2+} are accumulated as Ca^{2+} phosphate inside the matrix. It is not like TMRM which is being accumulated to equilibrium with the membrane potential.

Rizzuto: It is fair to conclude this in physiological conditions, but we have mitochondrial diseases and other pathological conditions in which this can vary. Making a titration of the severity of the respiratory deficiency could allow us to correlate it with the alteration in Ca^{2+} signalling. We haven't done this in detail apart from demonstrating that the respiratory deficiency decreases net uptake.

Jacobs: You showed one nice experiment, which in the light of what we heard yesterday supports the idea that the membrane potential is important. When you fragmented the mitochondrial network and prevented the fusion–fission cycle, you interpreted that to indicate that there is some kind of alteration in the propagation of the signal. To me it seemed that there were mitochondria dotted around the cell that simply can't respond.

Rizzuto: We interpreted it as a lack of diffusion, but you may be right: there could also be some bioenergetic explanation.

Larsson: In your paper you mainly looked at acute responses induced by various types of manipulations. We have looked in animal models with decreased oxidative phosphorylation capacity, both in the heart and skeletal muscle. In collaboration with a group from the Karolinska Institute we have examined Ca^{2+} signalling in isolated cardiomyocytes and skeletal muscle fibres. There are massive changes with smaller Ca^{2+} transients of shorter duration in respiratory chain deficient cardiomyocytes. We find a large secondary down-regulation of expression of many nuclear genes that encode proteins involved in Ca^{2+} handling, such as calsequestrin or SERCA2. These results suggest that reprogramming of gene expression causes aberrant Ca^{2+} metabolism.

Rizzuto: It is a difficult area because of the convergence of so many inputs. We don't know the molecular nature of the uniporter nor a reliable cell-permeant inhibitor, so we don't have a way to specifically affect the ability to take up Ca^{2+}. Any disease or bioenergetic manoeuvre we apply results in tens of changes in the cell. It is difficult to sort out specific Ca^{2+} signalling effects. We are now constructing a Ca^{2+} sponge located in mitochondria, so that we will be able to dampen significantly only the mitochondrial Ca^{2+} transients. We hope this will be a useful tool.

O'Rourke: You have to be careful in your conclusion that your PKC is modulating the uptake apparatus per se. It will modulate other things as well. For example, we have recently published a paper looking at fast mitochondrial Ca^{2+} uptake in heart cells (Maack et al 2006). It is highly Na^+ dependent. By modulating the efflux rate you affect the uptake rate as well. PKC is known to alter the Na^+/H^+ exchanger in the sarcolemma, among other effects.

Rizzuto: I am cautious by nature. We are forced to be reductionist. No doubt there are many other functions that have compensatory effects in the cell.

Nicholls: Can we focus on PKC and p66 for a moment?

Bernardi: This is an interesting observation. As you know the pore has been put online with p66 in the paper by Giorgio et al (2006). There is an additional piece of evidence which might be interesting here in a paper by Baldari and coworkers in Jurkat T cells (Pellegrini et al 2007). Ca^{2+}-dependent cell death in this system is strictly p66-dependent. She found that p66 phosphorylation itself is strictly Ca^{2+}-dependent. Pretreating with thapsigargin in Ca^{2+}-depleted cells prevents p66 phosphorylation. I see a lot of feedback going on here. If you need PKC to activate p66 you also need Ca^{2+} signalling in the cytosol. This rules out several Ca^{2+}-independent kinases such as ERK.

Rizzuto: You are right. We need to apply some protocols. Of course, we demonstrate that oxidative stress can activate PKCβ, but the key activator of PKCβ is Ca^{2+}.

Martinou: Is p66 phosphorylated at the plasma membrane?

Rizzuto: No, in the cytosol.

Martinou: Once it is phosphorylated, it enters the mitochondrion. So it must have a mitochondrial-targeting sequence.

Rizzuto: We are talking about just a fraction of p66. The majority remains in the cytosol where it plays a different role, i.e. it acts as a (relatively inefficient, compared to the other isoforms) growth factor adaptor.

Orrenius: Isn't it thought that once in the mitochondria, p66 is bound either to the TOM complex or to a heatshock protein, where it stays until it is released by a pro-apoptotic signal and can interact with cytochrome c?

Rizzuto: That is one scheme. We see that there is more p66 in mitochondria after PKC activation. The net amount of p66 in mitochondria is double when PKC is activated.

Jacobs: Given this nice hypothesis, and the benign mouse phenotype, wouldn't you expect p66 to be a tumour suppressor gene?

Rizzuto: It is not, which is surprising.

Jacobs: What about oxidative damage in the nuclear genome? Are the cells rendered susceptible to oxidative damage of DNA?

Rizzuto: This is the question that Pelicci is repeatedly asked since he published the *Nature* paper (Migliaccio et al 1999). The first question is why isn't this a tumour suppressor gene? He doesn't have the answer and neither do I. We need to understand how this apoptotic mechanism relates to immunosuppression of cancer cells. The other question is, why do we have it? Why hasn't it been selected against by evolution since we only see deleterious effects? The only piece of evidence for a negative effect is that the knockout mouse is less fertile. We don't understand why, but this could account for maintaining a potentially deleterious gene.

Martinou: I heard recently from Dr Pelicci at a meeting on apoptosis that they found these mice have problems in the brown adipose tissue. They are leaner. Dr Pelicci invoked the possibility that these mice eat less, and the lifespan extension could be consequent to caloric restriction.

Jacobs: Apoptosis is clearly important in many aspects of immune system function. One would also want to ask questions about whether the absence of this function made the mice susceptible to particular kinds of infectious challenge. It might be a specific effect.

Scorrano: The response of the immune system is very complicated, particularly if you are studying it at the whole animal level. Asking whether you have an immunological phenotype is too broad a question.

Jacobs: There are many ways you would want to ask this question. There is also the question of self-tolerance. If this balance were altered you would need specific assays to detect it.

Scorrano: You need to ask specific questions. For example, if Bcl-2 is expressed in T cells there are complex phenotypes, which depend on the dual effect of Bcl-2 on proliferation and apoptosis. If proliferation is suppressed on one side and apoptosis is suppressed on the other side, the net effect is that Bcl-2 doesn't have a major T cell phenotype. If we go to complex phenotypes such as immune system function, we should be extremely cautious and ask questions that are specific. I have a question: we know that hydrogen peroxide at 1 mM causes Ca^{2+} release from the ER and Ca^{2+} influx from the plasma membrane. Have you tried blocking calcineurin to see whether these changes still occur?

Rizzuto: We haven't, but we have done experiments replicating these effects with lower H_2O_2 concentrations. We still see the reduction of the Ca^{2+} transient.

Giulivi: You are studying this close interaction between IP_3 and mitochondria. The conclusion is that this close interaction results in Ca^{2+} that activates the dehydrogenases, produces ATP and favours the clearance of Ca^{2+} by the Ca^{2+}-dependent ATPases of the ER. You are looking at this interaction because it forces Ca^{2+} back to the ER. Is that what you are looking for?

Rizzuto: An important concept in Ca^{2+} signalling is that every time we deal with a channel we must consider that these channels are scaffolded in the plasma membrane as a signalling domain. The idea is that channels are not lost in the membrane, but are held together with the targets. This is an emerging theme in global Ca^{2+} signalling. This implies that regulators can be recruited to this domain and further participate in the regulation. The fact that this stimulates aerobic metabolism and activates the SERCA has been proven. What you say is right but it isn't directly related to this model. We find the SERCA in a different macromolecular complex. We still find it in these mitochondria-associated mitochondrial fractions, but not held together with IP_3 receptor and VDAC.

Giulivi: Do you think that this close interaction triggers the clearance of Ca^{2+}?

Rizzuto: This is a way to stimulate aerobic metabolism. Mitochondria produce ATP and this ATP is relevant for Ca^{2+} reuptake.

Nicholls: I want to ask a question for both of the last two talks. We have seen images in two dimensions but in reality we must think in three dimensions, with the complete 'sausage' of the mitochondrion. György shows us specific, almost tubular contacts between the ER and the mitochondria. Because we don't know anything about the Ca^{2+} uniporter, if we start with the observation that this is uniformly dotted around the inner membrane, then only a small proportion of the Ca^{2+} uniporters will be opposed to the ER. Many will be out looking at naked cytoplasm. What proportion of mitochondrial Ca^{2+} uptake in your different model systems occurs directly from release from the ER, or occurs from global bulk uptake from the cytoplasm?

Hajnóczky: In RBL-2H3 cells, the IP$_3$ receptor-mediated Ca^{2+} release appears to maximally activate the mitochondrial Ca^{2+} uptake. In the context of the local Ca^{2+} control model this result suggests that the distribution of the mitochondrial uptake sites (uniporter and VDACs) allows all of them to interact with an IP$_3$ receptor. The IP$_3$ receptors form clusters in the ER membrane. We speculate that the Ca^{2+} uptake sites may show an inhomogeneous distribution in the mitochondrial membranes. In the RBL-2H3 cells, the mitochondria can take up 40% of the Ca^{2+} released from the ER. This seems to be the high end. In some other cell types, mitochondria would accumulate only a few percent of the Ca^{2+} released through the IP$_3$ receptors.

Rizzuto: Some VDAC is there and some is diffused. The same is true for the IP$_3$ receptor. For the uniporter it is difficult to understand how much is clustered, since its molecular identity is still elusive. In principle, however, given that its Ca^{2+} response is steep, the coupled uniporter is the relevant one, even if it is only a fraction. There is evidence indicating that the macromolecular complexes bring together a signalling domain, through the work of Hajnóczky and Duchen. It is clear that mitochondrial Ca^{2+} uptake has a fundamental role in deciding the opening probability of the IP$_3$ receptor. If these IP$_3$ receptors were scattered through the whole ER, it would be irrelevant if a fraction of the IP$_3$ receptors are modulated by mitochondria. Their data show that the IP$_3$ receptors are modulated by mitochondria, which implies that the signalling proteins come together. If the IP$_3$ receptor comes together it makes a lot of sense that the uniporter is there.

Hajnóczky: We have talked about the IP$_3$ receptor, but in other cell types, the ryanodine receptors also form local interactions with the mitochondria to support the Ca^{2+} signal propagation from the sarcoplasmic reticulum to the mitochondria.

Orrenius: Importantly, VDAC is the most abundant protein in the outer mitochondrial membrane. Further, formation of a complex between VDAC and the IP$_3$ receptor may also be of interest in view of the observation by Snyder and colleagues (Boehning et al 2003) that cytochrome c can interact with the IP$_3$ receptor to delay closing of the receptor by the increasing cytosolic Ca^{2+} concentration during apoptosis. This might also provide a direct mechanism for filling the mitochondria with Ca^{2+} under apoptotic conditions.

Parekh: We and others have been patch clamping mitoplasts. It is hard to find a single uniporter channel: the density seems quite low. But when we find one we discover that they have a relatively high single channel Ca^{2+} conductance and a high open probability. We don't need many of them to get rapid mitochondrial Ca^{2+} uptake. In Rizzuto's model, judicious location of the uniporter and IP$_3$ receptor by VDAC would result in effective coupling.

ER/MITOCHONDRIAL CROSS-TALK 137

Nicholls: Can you make an estimate of how many uniporters there are per mitochondrion?

Parekh: The open probability seems quite high, so perhaps 50 per mitoplast, which is around 1 µm in diameter.

Nicholls: This could explain why we have looked for 40 years and not found any.

O'Rourke: I wanted to follow up on the IP$_3$ receptor binding domain. Have you looked at the K_m for ADP stimulation of respiration when you overexpress? This would give an idea of whether the VDAC conductance to nucleotides is altered.

Rizzuto: That's a good point. We haven't.

Shirihai: According to what you showed, the mitochondrion that falls into a pool of Ca^{2+} is doomed to stay there forever, and perhaps this would contribute to the initiation of apoptosis.

Nicholls: So you could get Ca^{2+} overload in specific mitochondria.

Hajnóczky: In the normal situation the recruitment of more mitochondria would work the opposite way. Mitochondria will provide the ATP and help the ER to reaccumulate the Ca^{2+}. They will also contribute to the buffering of the Ca^{2+}. If the resident mitochondria are not enough to support the Ca^{2+} handling, more mitochondria will be retained in the high Ca^{2+} zone to help out. Under excessive Ca^{2+} release or Ca^{2+} load conditions, the capacity may be worn down. But the Ca^{2+}-dependent control of the mitochondrial distribution seems to decrease the chance for the overloading.

Shirihai: Would you expect the Ca^{2+} to reduce the fusion capacity of these mitochondria?

Hajnóczky: For the period the mitochondria spend in the vicinity of the Ca^{2+} release, the fusion activity is probably reduced.

Rizzuto: With regard to the contacts, I am not discussing what creates the contacts. Once the contacts between the two organelles are made, this assures that the channels are put together and can talk to each other. But I don't think that the number of contacts is determined by the chaperone. This is a modulator of the function. Once you have created this domain, it allows the two systems to talk to each other.

Nicholls: You are uniformly increasing Ca^{2+} in your cells, and the mitochondria stop. Can one devise an experiment where there is focal stimulation of one pole of a mitochondrion?

Reynolds: Peter Hollenbeck published an interesting experiment in which he was looking at mitochondria travelling up and down a DRG axon. He took a bead of NGF, put a spot on the axon and then looked at the probability that mitochondria would stop. They pause on the way down. It is not quite the same thing as attraction, but they have a tendency to linger when they get there. This addresses the issue of activating the process that promotes docking. You raised the possibility of a hypothetical docking protein. You say you don't think the adaptor proteins are making the mitochondria stop, but perhaps they do.

Rizzuto: Not this adaptor protein. There may be others that do.

Shirihai: Is that what would keep mitochondria in active synapses.

Nicholls: That would be the hope. We always talk about an axon with a synapse at one end, which would be the neuromuscular junction, but real CNS synapses have hundreds of varicosities along them. Do the mitochondria derail themselves at particular synapses that are being active, saying 'hey we are needed here' because the Ca^{2+} is elevated?

Scorrano: There is a paper by Antonella Viola's group (Campello et al 2006). We collaborated with them. She showed that mitochondria were recruited at the uropods of migrating lymphocytes. This recruitment is separate from the recruitment of the ER and depends on the activation of the fission machinery. If you block fragmentation of the mitochondria, you block polarization of the T cell. If you induce fragmentation of the mitochondria by itself, this drives polarization of the T cell. This is Ca^{2+} independent.

Nicholls: Is there any relationship between the fission/fusion state and where the mitochondria will end up? Why are presynaptic mitochondria smaller than somatic mitochondria?

Youle: The inherited diseases of mitochondrial fission and fusion proteins are neuronal, such as CMT2A.

Nicholls: You showed some nice fission and fusion data. Does anyone want to comment?

Shirihai: These data are nice. I would like to comment on the use of the term 'kiss and run' here. Usually we use the term 'kiss and run' where there is a sharing of solution content but not where sharing of membranes is found. Here there is clearly a sharing of membrane components; although the kinetics are slower. If this is a 'kiss and run' event, it is a definitely a slow French kiss followed by a slow-motion run.

Duchen: One of the things you showed was that the more depolarized mitochondria were less likely to fuse later. Is this because they might be less likely to move, or could it be because they are Ca^{2+} loaded?

Shirihai: It is possible that the reason the depolarized mitochondria are less likely to fuse is their association with the cytoskeleton. We need to examine this.

Hajnóczky: With regard to the Ca^{2+} dependence, we have detected fusion events in Ca^{2+}-depleted cells. Thus a Ca^{2+} elevation in the cytosol or organelles does not seem to be required for the fusion event. Regarding the membrane potential dependence of the fusion events, the view has been that the membrane potential is critical, at least for the inner membrane fusion. Recording of the membrane potential simultaneously with the fusion events gives a hint that the fusion capacity is not lost at the same time as the fall in membrane potential. If the membrane potential is collapsed by the addition of FCCP, there are still fusion events in the next few minutes. Perhaps it is not the membrane potential per se that is needed for the mitochondrial fusion, but the membrane potential loss is probably converted to a chemical signal such as a cleavage of a protein, which prevents further fusion events. Perhaps this mechanism contributes to the selection of the pool that then undergoes autophagy.

Halestrap: How much is exchanged in a 'kiss and run'? We know a lot about the structure of a mitochondrion. Many of the proteins are in a sort of matrix gel. They are hardly free. When you add an indicator like an extraneous protein, this may well move much better. But how many of the normal proteins move?

Schon: An experiment was done by Nonaka in which he fused ρ^0 cells with wild-types and looked for a recovery of cytochrome oxidase in the fusion product. It is remarkably fast.

Halestrap: In each 'kiss and run' it may be a relatively small amount that is exchanged.

Reynolds: In terms of the magnitude of the structures we are looking at here, how much DNA is there in these pieces of mitochondria that are $2\mu m$ long when they go into fission or fusion? Is every daughter going to have a piece of DNA?

Schon: Yes. It's amazing. If cells are stained with MitoTracker and you look for nucleoids, no matter what the size of the MitoTracker-positive object, it will have at least one nucleoid in it. We have to be careful because it could be that only nucleoid-containing mitochondria become MitoTracker positive, but it is a remarkable correlation.

References

Boehning D, Patterson RL, Sedaghat L, Glebova NO, Kurosaki T, Snyder SH 2003 Cytochrome c binds to inositol (1,4,5) tris-phosphate receptors, amplifying calcium-dependent apoptosis. Nat Cell Biol 5:1051–1061

Campello S, Lacalle RA, Bettella M, Manes S, Scorrano L, Viola A 2006 Orchestration of lymphocyte chemotaxis by mitochondrial dynamics. J Exp Med 203:2879–2886

Giorgio M, Migliaccio E, Orsini F et al 2005 Electron transfer between cytochrome c and p66Shc generates reactive oxygen species that trigger mitochondrial apoptosis. Cell 122:221–233

Maack C, Cortassa S, Aon MA, Ganesan AN, Liu T, O'Rourke B 2006 Elevated cytosolic Na+ decreases mitochondrial Ca2+ uptake during excitation-contraction coupling and impairs energetic adaptation in cardiac myocytes. Circ Res 99:172–182

Migliaccio E, Giorgio M, Mele S et al 1999 The p66shc adaptor protein controls oxidative stress response and life span in mammals. Nature 402:309–313

Pellegrini M, Finetti F, Petronilli V et al 2007 p66SHC promotes T cell apoptosis by inducing mitochondrial dysfunction and impaired Ca2+ homeostasis. Cell Death Differ 14:338–347

Mitochondrial ion channels in cardiac function and dysfunction

Brian O'Rourke, Sonia Cortassa, Fadi Akar and Miguel Aon

The Johns Hopkins University, Institute of Molecular Cardiobiology, Division of Cardiology, Department of Medicine, Baltimore, MD 21205, USA

Abstract. The study of mitochondrial physiology continues to provide new and surprising insights into how this organelle participates in the integration of cellular activities, far beyond the traditional view of the mitochondrion in energy transduction. Emerging evidence indicates that mitochondria are a centre of organization of numerous signalling pathways and are a cellular target that undergoes vast modification during both the acute and chronic phases of disease development and ageing. In this context, it is also important to understand the spatial and temporal organization of mitochondrial function and how this might influence the cell's response to stress. Here, we present evidence supporting the hypothesis that mitochondria from heart cells act as a network of coupled oscillators, capable of producing frequency- and/or amplitude-encoded reactive oxygen species (ROS) signals under physiological conditions. This intrinsic property of the mitochondria can lead to a mitochondrial 'critical' state, i.e. an emergent macroscopic response manifested as complete collapse or synchronized oscillation in the mitochondrial network under stress. The large amplitude depolarizations of $\Delta\Psi_m$ and bursts of ROS have widespread effects on all subsystems of the cell including energy-sensitive ion channels in the plasma membrane, producing an effect that scales to cause organ level electrical and contractile dysfunction. Mitochondrial ion channels appear to play a key role in the mechanism of this non-linear network phenomenon and hence are an important target for potential therapeutic intervention.

2007 Mitochondrial biology: new perspectives. Wiley, Chichester (Novartis Foundation Symposium 287) p 140–156

Multiple non-linear control interactions govern mitochondrial oxidative phosphorylation, allowing the mitochondrial to adjust energy production to large changes in demand, while keeping the many positive and negative feedback loops in check. Thus, there must be a balance between a flexible response to the intracellular environment and a robust resistance to instability, both from the point of view of matching energy supply with demand, and to limit the toxic effects of the by-products of metabolism, including reactive oxygen species (ROS). As is often the case for highly responsive control systems, the inherent non-linear properties of a system can sometimes result in chaotic or unstable behaviour if the system is

CARDIAC FUNCTION 141

stressed beyond its normal range of behaviour. A central point of this paper is that the mitochondrial network of the heart cell has a critical threshold of oxidative stress that constitutes a breakpoint between physiological and pathophysiological domains of behaviour.

Several features of mitochondria are vital to understanding why function can fail catastrophically under stress. First, there is a large protonmotive force across the inner membrane, and the electrical potential ($\Delta\Psi_m$) makes up the bulk of this force. Second, there is a low permeability to ions across the inner membrane. Third, the mitochondrial matrix is a limited space bounded by a double membrane, therefore, there are physical constraints on ion and osmolyte movements that might influence function. Hence, there is a large electrochemical gradient for cation and anion movements across the inner membrane, but they are kept in check by relatively low permeabilities and the need for charge and volume balance to be maintained. It is easy to understand then, that if a channel opens in the inner membrane, it will dissipate energy, the magnitude of the effect being related to the conductance and specificity of the ion channel. While many mitochondrial ion channels have been proposed to be present in the inner membrane from classical swelling experiments and electrophysiological recordings, we are still struggling to characterize their structures and physiological functions (O'Rourke 2006). This area represents an important goal for mitochondrial research in general.

From the therapeutic perspective, particularly with reference to ischaemia–reperfusion injury in the heart and brain, it is important to note that mitochondrial ion channels appear to play a role in both protection against injury and as mediators of injury. An increase in mitochondrial K^+ flux, for example, by pretreating hearts with K, ATP or K_{Ca} channel opener compounds, which are thought to activate mitochondrial K_{ATP} (mitoK_{ATP}) or K_{Ca} channels (mitoK_{Ca}), respectively, has been found to significantly decrease infarct size and improve recovery after ischaemia–reperfusion. On the other side of the equation, blocking mitochondrial ion channels, such as the permeability transition pore (PTP), or the inner membrane anion channel (IMAC), can prevent the loss of mitochondrial function that is a prelude to necrotic or apoptotic cell death.

Based on the novel concepts described above, in the present paper we will summarize some of the emerging ideas we have about the spatiotemporal organization of mitochondria and the role of mitochondrial ion channels in the critical transition between physiology and pathophysiology.

Mitochondrial ion channels involved in cell stress responses

While the concept of mitochondrial ion channels (or uniporters) has been around for quite some time, renewed interest has been generated as their primary roles as determinants of cell life and death have been revealed (Aon et al 2006a, O'Rourke et al 2005). Our primary focus has been on those channels involved in either protecting cardiac cells from injury, or causing ischaemia and reperfusion injury. In this context, we have employed a variety of methods to identify and characterize which ion channels may be present in isolated mitochondria and in intact cells.

K^+ channel opener compounds, including diazoxide, nicorandil and others, can protect heart cells from ischaemic or oxidative stress through a mechanism which we believe involves the opening of specific mitoK$_{ATP}$ channels on the inner membrane (O'Rourke 2004). Similarly, studies have revealed a second class of K^+ channel on the mitochondrial inner membrane (mitoK$_{Ca}$), resembling the Ca^{2+}-activated K^+ channel of the plasmalemma of certain cell types. Selective openers of the K$_{Ca}$ channel activated this channel (e.g. NS-1619) and it was inhibited by specific toxins (Xu et al 2002). MitoK$_{Ca}$ activation conferred protection against ischaemia-reperfusion injury, which was prevented by K$_{Ca}$ inhibitors, providing further support for the idea that increased K^+ flux can protect the myocardium.

The role of mitochondrial Ca^{2+} uptake and ROS accumulation, and the eventual activation of the PTP, has been shown to play a prominent role in reperfusion injury (Halestrap et al 2004). While the PTP is apparently triggered during reperfusion, other mitochondrial ion channels may be activated during ischaemia, to cause loss of $\Delta\Psi_m$. In particular, our recent studies have focused on the activation of IMAC, an outwardly rectifying anion channel that is modulated by mitochondrial benzodiazepine receptor (mBzR) ligands. Our current computational and experimental studies place IMAC at the centre of a mitochondrial ROS-induced ROS release mechanism that underlies the oscillatory properties of the mitochondrial network.

Mitochondria as a network of coupled oscillators

Mitochondrial criticality and large amplitude oscillation under pathological conditions

Our early studies of metabolic stress in isolated cardiac cells, in the form of substrate deprivation, revealed that energy-sensitive K^+ channels in the sarcolemmal membrane can be activated spontaneously in an oscillatory manner (O'Rourke et al 1994). These K$_{ATP}$ current oscillations were closely associated with whole cell metabolic oscillations in the NADH redox pool. Modulation of the cellular action potential by these metabolic oscillations led us to hypothesize that these oscillations could result in arrhythmias if present in the heart after ischaemia–reperfusion. Subsequently, we identified the mitochondria as the source of the oscillations and observed that they involved the synchronized depolarization of the mitochondrial network of almost the entire heart cell (Romashko et al 1998). Based on the observation that $\Delta\Psi_m$ depolarization could either occur in individual mitochondria, clusters of mitochondria, or in the whole network, we suggested that the mitochondria may represent a network of coupled oscillators (Romashko et al 1998).

Since the mitochondrial network of the cardiac cell consists of thousands of mitochondria packed between the myofilaments in an ordered three-dimensional

CARDIAC FUNCTION 143

array (or lattice), we tested whether mitochondria behaved independently or were synchronized, and what factors might be responsible for intermitochondrial communication. High resolution, two-photon laser scanning fluorescence imaging was used to track $\Delta\Psi_m$ (with TMRE), ROS production (with matrix localized derivatives of dichlorofluorescein, a reporter of H_2O_2), and NADH (native autofluorescence) simultaneously, and a single localized high intensity laser flash was applied in a small volume of the mitochondrial network (an $8\,\mu m \times 8\,\mu m$ square, approximately $1\,\mu m$ deep). This flash rapidly depolarizes mitochondria and generates ROS in the flashed region, but no obvious effect on the remainder of the mitochondrial network was initially produced (Figs 1A and B). However, signs of spreading oxi-

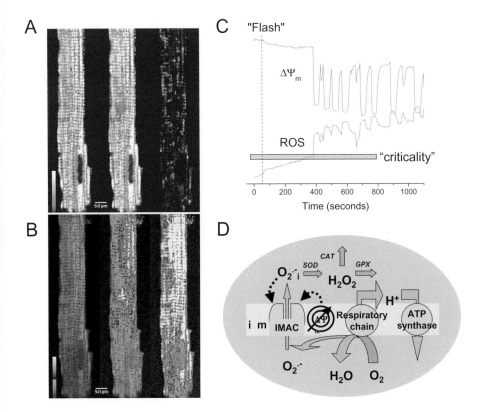

FIG. 1. Mitochondrial network depolarization after a local laser flash. (A) Upper panels: $\Delta\Psi_m$ signal before the flash, close to criticality, and after global depolarization. (B) ROS signal before the flash, close to criticality, and after global depolarization. (C) The number of mitochondria with ROS above threshold increases to ~60% at criticality just prior to global depolarization and limit cycle oscillation of the network (adapted from Aon et al 2004). (D) Conceptual and computational model of the mechanism of the ROS-dependent mitochondrial oscillator (adapted from Cortassa et al 2004).

dative stress are observed: the mitochondrial matrix ROS signal increases in more and more mitochondria over the next 1–2 minutes, although $\Delta\Psi_m$ is unchanged. When a certain proportion of mitochondria (60%) in the 2D field show a ROS signal increase by 20% or more, then a global synchronized depolarization of the mitochondrial network occurs. A limit cycle oscillation in the system is then triggered, with a reproducible period of roughly 1.5 min (Aon et al 2003) (Fig. 1C). These oscillations can be prevented or even blocked acutely, by inhibiting IMAC with mBzR ligands such as 4'Cl-diazepam (Ro5 4864) or PK11195, or by DIDS, a non-specific anion transport inhibitor. Importantly, all interventions that would be expected to inhibit superoxide production by the Q cycle of complex III were also effective at stabilizing $\Delta\Psi_m$ after flash-induced oscillations were triggered (Aon et al 2003).

We referred to the state just prior to global mitochondrial depolarization as the point of 'mitochondrial criticality' (Fig. 1C). The behaviour of the mitochondrial network at criticality was consistent with the formation of a percolation or spanning cluster across the network (Aon et al 2004). This theory describes how neighbouring elements (mitochondria) in the lattice not only influence the local response, but also contribute to an emergent macroscopic response of the entire network. When a critical mass of mitochondria are at the percolation threshold, a small perturbation in any of the mitochondria of the spanning cluster can result in the whole-cell state transition. This characterization helps to account for the observation that the pattern of $\Delta\Psi_m$ depolarization is reproducible from one cycle of oscillation to the next (some mitochondria even remain polarized in the midst of widespread depolarization around them) and the lack of any specific centre of origin of each depolarization (as is often observed for spontaneous Ca^{2+} oscillations in heart cells).

Uniquely, the ability to trigger reproducible cell-wide synchronized autonomous oscillations of $\Delta\Psi_m$, NADH, ROS (and also reduced glutathione) in cardiac myocytes using a highly localized oxidative trigger enabled us to explore the mechanism and the functional implications of this spatiotemporal dynamic response in unprecedented detail. The oscillatory behaviour could also be reproduced in a mathematical model of a single mitochondrion, which incorporates ROS production by the electron transport chain, a ROS-activated mitochondrial channel (IMAC), and extramitochondrial ROS scavenging (Cortassa et al 2004) (Fig. 1D). In this model, ROS produced by the electron transport chain accumulates to a threshold level, triggering the opening of IMAC in a positive feedback loop. IMAC activation is terminated by a reduction in ROS at the activator site of the channel as a result of membrane depolarization (decreasing ROS production and efflux from the mitochondrial matrix) and ROS scavenging by the antioxidant enzymes. Thus, the system acts as a relaxation oscillator, in which a controlling factor builds up to a critical point, and then a rapid change is triggered, with the process then

CARDIAC FUNCTION 145

repeating in a stereotypical pattern. These events, which the model suggests can occur in single mitochondria, are transmitted to the whole mitochondrial network through local ROS-dependent interactions among mitochondria.

Weakly coupled oscillations revealed by long range temporal correlations of $\Delta\Psi_m$

In an interesting example of how computational modelling can enhance interpretation of data as well as suggest new ideas to test experimentally, we observed very soon after exploring the parameter space of the mitochondrial oscillator that the frequency and amplitude of mitochondrial oscillation could span a remarkably wide range, from milliseconds to multiple hours, and µV to 100 mV (Fig. 2). We hypothesized that one of two possibilities for the transition between stable physi-

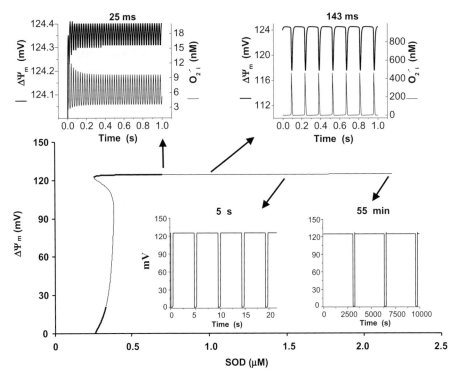

FIG. 2. Frequency and amplitude range of the mitochondrial oscillator. Variation of a single parameter (the superoxide dismutase activity) in a computational model of the ROS-dependent mitochondrial oscillator produces a wide range of behaviours from low amplitude fast oscillation (25 ms) to slow, large amplitude oscillations in $\Delta\Psi_m$ (55 min). The rapid phases of $\Delta\Psi_m$ depolarization (uncoupling) are accompanied by small pulses (fast domain) or bursts (slow domain) of superoxide release from the matrix to the intermembrane space (see Cortassa et al 2004 for details).

ological behaviour and unstable pathophysiological behaviour were possible. Either the system was (i) undergoing a bifurcation in the dynamics from a stable steady state to limit cycle oscillation under stress, or (ii) the mitochondria were oscillating with high frequency and low amplitude under 'normal' conditions and there was an increase in the synchronization and coupling of the oscillators in a dominant low frequency, high amplitude mode at the critical point.

To determine if there was evidence of correlated noise due to the oscillatory dynamics of mitochondrial energetics under physiological conditions, we recorded long time series of $\Delta\Psi_m$ at a fast frame rate (~100 ms) and applied two methods to look for long range statistical correlations in the data (Fig. 3). The first method was to apply relative dispersion analysis (RDA), a form of detrended fluctutation analysis, to the data (Aon et al 2006b). The relative dispersion (standard deviation/mean) was determined using increasing window sizes for aggregating the data and the slope of the log of RD versus the log of the aggregation number was determined. In this analysis, completely random fluctuations yield a slope (or fractal dimension, Df) of 1.5. Notably, Df was close to 1.0 for $\Delta\Psi_m$ fluctuations under both physiological and pathophysiological conditions. This indicates that there is long term memory in the system, that is, the current state of $\Delta\Psi_m$ is correlated with $\Delta\Psi_m$ in the past, over several time scales, and this observation is consistent with the behaviour of coupled oscillators.

The second method was to use power spectral analysis (PSA) to determine if the noise was correlated. The slope (β) of the log-log plot of power vs. frequency demonstrated that $\Delta\Psi_m$ follows a power-law dependence according to $f^{-\beta}$, with $\beta = 1.74$. Again, this slope is distinct from random noise ($\beta = 0$), and indicates that there is a broad spectrum of frequency components spanning several orders of magnitude underlying the apparently 'stable' $\Delta\Psi_m$ observed under physiological conditions.

Using the same respiratory or IMAC inhibitors employed to investigate the mechanism of the mitochondrial oscillator, we demonstrated that the same interventions that abruptly stopped the large amplitude oscillations also decreased the extent of the correlation in the PSA, particularly in the high frequency part of the spectrum. We therefore propose that the mitochondria are normally functioning as a collection of weakly coupled oscillators that, under stress, can become strongly coupled by ROS to produce a dominant slow, large amplitude oscillation in the network. The latter condition is characterized by large bursts of ROS production during the uncoupling phase of the cycle, which will eventually overwhelm the antioxidant capacity of the cell and lead to cell death.

Mitochondrial criticality as the origin of contractile and electrical dysfunction during ischaemia and reperfusion

The primary function of the cardiac cell, excitation–contraction coupling, is intimately coupled to the energetic status. In the oscillatory phenomenon described

CARDIAC FUNCTION

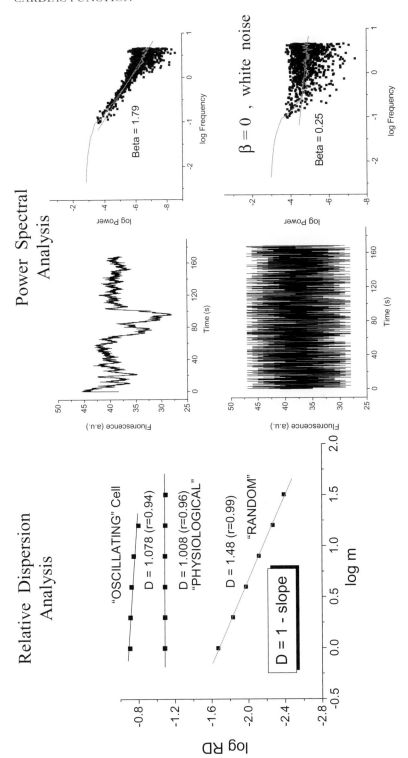

FIG. 3. Relative dispersion analysis and power spectral analysis of $\Delta\Psi_m$ under physiological and pathophysiological (oscillating cell) conditions. Left panel: RDA demonstrates that under both conditions, long-range correlations in the signal are present, distinct from random noise. Similarly, the power spectrum shows a broad frequency dependence dropping off with a $1/f^{1.79}$ dependence, in this example. Both methods suggest that mitochondria are organized as a collection of weakly coupled oscillators under normal conditions (see Aon et al 2006 for details).

above, the repeating cycles of mitochondrial $\Delta\Psi_m$ depolarization and repolarization allow us to examine the phase relationships between $\Delta\Psi_m$, ROS bursts, and their effects on cellular action potentials and Ca^{2+} transients. The rapid phase of mitochondrial uncoupling was closely correlated with the activation of energy-sensing K_{ATP} channels of the sarcolemmal membrane (Aon et al 2003, O'Rourke et al 1994), and also with the suppression of intracellular Ca^{2+} release (O'Rourke et al 1994). This demonstrates that mitochondrial criticality scales to produce global dysfunction at the level of the whole cardiac cell. The modulation of cellular electrical excitability is particularly relevant to the generation of cardiac arrhythmias, because dispersion of repolarization of the myocardium is known to be a main factor contributing to the development of reentrant circuits.

In recent studies, we have provided evidence that failure of the mitochondrial network not only scales to the level of the whole cell, but also underlies global electrical dysfunction in the whole heart during ischaemia and reperfusion (Akar et al 2005). Isolated-perfused guinea-pig hearts were subjected to 30 minutes of global ischaemia followed by reperfusion while epicardial electrical activity was followed using a multichannel optical mapping system. In more than 90% of the control hearts, ventricular tachycardia and fibrillation was induced upon reperfusion. Exposure of the hearts to 4′Cl-diazepam, which was shown to stabilize both $\Delta\Psi_m$ and the action potential duration of cardiac cells undergoing oscillations after a laser-induced flash, completely eliminated post-ischaemic arrhythmias (Akar et al 2005). We proposed that clusters of cells that have reached mitochondrial criticality in the heart during ischaemia constitute metabolic current sinks that will impede electrical propagation due to their high K_{ATP} conductance. These regions could either be completely unexcitable, or could have marked spatiotemporal heterogeneity of the action potential duration that promotes reentry.

Studies are currently underway to determine if regional and heterogeneous conduction slowing correlates with regional heterogeneous mitochondrial depolarization. Preliminary evidence indicates that IMAC inhibition protects against the loss of $\Delta\Psi_m$ during ischaemia, prevents heterogeneity of $\Delta\Psi_m$ upon reperfusion, and preserves systolic and diastolic contractile function (data not shown).

Outstanding questions/new perspectives

Mitochondrial ROS production as a frequency and amplitude encoded signalling system

While the contribution of mitochondrial ROS to cell injury and death has been well established, the physiological relevance of ROS-dependent signalling is just beginning to be realized. A clear example is the participation of ROS in the activation of cardioprotective pathways (Oldenburg et al 2002). The presence of ROS scavengers can completely prevent ischaemic or pharmacological preconditioning.

CARDIAC FUNCTION

Recent evidence has also implicated mitochondrial ROS in the activation of the HIF1α transcriptional response during hypoxia (Waypa et al 2006).

The finding that the mitochondrial network behaves as a network of coupled oscillators immediately suggests that this property might constitute a mechanism for fine modulation of ROS signals, to be decoded by redox-sensitive transcription factors, ROS-sensitive kinase/phosphatase systems, or indirectly, for example, through effects on Ca^{2+} handling. Analogies exist with other frequency-modulated systems, such as the Ca^{2+}/calmodulin dependent kinase-mediated phosphorylation pathway (Hanson et al 1994) and hormone stimulated Ca^{2+} signalling in nonexcitable cells (Hajnoczky et al 1995). In fact, the latter has recently been shown to be dependent on mitochondrial ROS generation (Camello-Almaraz et al 2006). It has been argued that frequency-encoded systems permit the cell to more easily detect a signal above background than an amplitude modulated response.

In terms of the physiological state, it remains to be determined if and how a broad spectrum of low amplitude $\Delta\Psi_m$ oscillations, with presumably small amplitude ROS oscillations, is decoded by the rest of the cell to produce a response. We suggest that the spectral power will change as a consequence of normal cell function, for example, in response to increased workload, a change in nutrient status, or a change in second-messenger mediated signalling pathways. Much more work will be needed to establish the cause and effect relationship between such factors, but the linkage between respiration and ROS production could be a direct reporter of the global cell status, and thus uniquely positioned to bring about a response.

Of course, the positive relationship between respiration and ROS production that we have observed during mitochondrial $\Delta\Psi_m$ oscillation directly contradicts some models of ROS production by the respiratory chain, as we have previously pointed out (O'Rourke et al 2006), but is consistent with ROS production from complex III (Turrens et al 1985). So the basic question of whether mitochondrial ROS production increases or decreases with an increase in respiratory rate in intact cells needs to be resolved in future studies. It is perfectly reasonable to assume that ROS can arise from different points in the respiratory chain and may be generated on different sides of the inner membrane under different conditions, for example during hypoxia versus during reoxygenation.

How does the morphology of the mitochondrial network influence cell function?

Clearly, in the cardiac cell, the mitochondrial network is highly organized, almost as a cubic array. This arrangement is probably necessary to serve the massive energy demands of muscle contraction, but it also maximizes neighbour–neighbour interaction in the network. The mitochondria are not only situated close to the myofilaments, the main sites of ATP hydrolysis, but also butt directly up to the Ca^{2+} release sites of the dyad, suggesting that local microdomain interactions may be important. Mitochondria may both shape and respond to local Ca^{2+} signals (Maack et al 2006), but may also modulate Ca^{2+} release by controlling the local redox environment or local ATP/ADP ratio, considering the fact that every one of the proteins involved in Ca^{2+} handling are sensitive to these factors.

It will therefore be interesting to investigate whether ROS-dependent mitochondrial network signalling plays a role in other cell types in which the mitochondria are distributed either as filamentous tangles, long strands, or as punctate spots. Diffusion distances should play a major role in the synchronization of these networks.

Mitochondrial ion channels: How selective are they? What are they?

A major source of controversy and frustration in the field has been the lack of molecular structure for all but a few mitochondrial ion channels (e.g. UCP, VDAC). Some of the difficulty in identifying and characterizing these channels is that they are likely to be present in low abundance, in order for the mitochondria to preserve the low permeability so essential to chemiosmotic energy transduction. Another challenge is to find a suitable method for assaying their expression. However, the ample and varied evidence that selective ion channels are present gives us hope that the hard work currently being carried out in several laboratories around the world will eventually bear fruit. Achievement of this goal will not only resolve some of the ongoing arguments, but will provide the basis for a new phase of mitochondrial research, the molecular dissection of mitochondrial ion channel function.

Conclusion

In summary, we have emphasized that the spatial and temporal organization of mitochondria is crucial to understanding the behaviour of the mitochondrial network in intact cells. In heart cells, mitochondria appear to be organized as a collection of weakly coupled oscillators that, under stress, can become strongly coupled by local ROS-induced ROS release. Synchronized oscillation of the mitochondrial network produces dramatic effects on the whole cell electrical and Ca^{2+} handling functions and these can scale to the whole organ to induce fatal arrhythmias and impaired contractile function. A particular mitochondrial ion channel, IMAC, appears to be a key player in this cascade of failures, while other mitochondrial ion channels may protect against mitochondrial dysfunction. Identification and molecular characterization of these mitochondrial ion channels will be necessary to gain a deeper understanding of the regulation of oxidative phosphorylation and intracellular signalling. It will also be important for developing novel mitochondrially targeted therapies for cardiovascular diseases, metabolic syndrome, neurodegeneration and ageing in the years to come.

References

Akar FG, Aon MA, Tomaselli GF, O'Rourke B 2005 The mitochondrial origin of postischemic arrhythmias. J Clin Invest 115:3527–3535
Aon MA, Cortassa S, Marban E, O'Rourke B 2003 Synchronized whole cell oscillations in mitochondrial metabolism triggered by a local release of reactive oxygen species in cardiac myocytes. J Biol Chem 278:44735–44744
Aon MA, Cortassa S, O'Rourke B 2004 Percolation and criticality in a mitochondrial network. Proc Natl Acad Sci USA 101:4447–4452
Aon MA, Cortassa S, Akar FG, O'Rourke B 2006a Mitochondrial criticality: a new concept at the turning point of life or death. Biochim Biophys Acta 1762:232–240
Aon MA, Cortassa SC, O'Rourke B 2006b The fundamental organization of cardiac mitochondria as a network of coupled oscillators. Biophys J 91:4317–4327
Camello-Almaraz MC, Pozo MJ, Murphy MP, Camello PJ 2006 Mitochondrial production of oxidants is necessary for physiological calcium oscillations. J Cell Physiol 206: 487–494
Cortassa S, Aon MA, Winslow RL, O'Rourke B 2004 A mitochondrial oscillator dependent on reactive oxygen species. Biophys J 87:2060–2073
Hajnoczky G, Robb-Gaspers LD, Seitz MB, Thomas AP 1995 Decoding of cytosolic calcium oscillations in the mitochondria. Cell 82:415–424
Halestrap AP, Clarke SJ, Javadov SA 2004 Mitochondrial permeability transition pore opening during myocardial reperfusion—a target for cardioprotection. Cardiovasc Res 61:372–385
Hanson PI, Meyer T, Stryer L, Schulman H 1994 Dual role of calmodulin in autophosphorylation of multifunctional CaM kinase may underlie decoding of calcium signals. Neuron 12:943–956
Maack C, Cortassa S, Aon MA, Ganesan AN, Liu T, O'Rourke B 2006 Elevated cytosolic Na^+ decreases mitochondrial Ca^{2+} uptake during excitation-contraction coupling and impairs energetic adaptation in cardiac myocytes. Circ Res 99:172–182
Oldenburg O, Cohen MV, Yellon DM, Downey JM 2002 Mitochondrial K(ATP) channels: role in cardioprotection. Cardiovasc Res 55:429–437
O'Rourke B 2004 Evidence for mitochondrial K^+ channels and their role in cardioprotection. Circ Res 94:420–432
O'Rourke B 2006 Mitochondrial ion channels. Annu Rev Physiol 69:19–49
O'Rourke B, Ramza BM, Marban E 1994 Oscillations of membrane current and excitability driven by metabolic oscillations in heart cells. Science 265:962–966
O'Rourke B, Cortassa S, Aon MA 2005 Mitochondrial ion channels: gatekeepers of life and death. Physiology (Bethesda) 20:303–315
Romashko DN, Marban E, O'Rourke B 1998 Subcellular metabolic transients and mitochondrial redox waves in heart cells. Proc Natl Acad Sci USA 95:1618–1623
Turrens JF, Alexandre A, Lehninger AL 1985 Ubisemiquinone is the electron donor for superoxide formation by complex III of heart mitochondria. Arch Biochem Biophys 237: 408–414
Waypa GB, Guzy R, Mungai PT et al 2006 Increases in mitochondrial reactive oxygen species trigger hypoxia-induced calcium responses in pulmonary artery smooth muscle cells. Circ Res 99:970–978
Xu W, Liu Y, Wang S et al 2002 Cytoprotective role of Ca^{2+}-activated K^+ channels in the cardiac inner mitochondrial membrane Science 298:1029–1033

DISCUSSION

Nicholls: I'm missing something here. Why do the mitochondria depolarize and how does this set up an oscillation?

O'Rourke: Essentially, what we have is a relaxation oscillator. There is a build up of ROS that reaches a critical threshold level and increases the opening probability of the inner membrane channel. This leads to energy dissipation and mitochondrial depolarization.

Nicholls: Why would opening of an inner membrane anion channel collapse the mitochondrial membrane potential?

O'Rourke: The idea would be that this channel has an equilibrium potential very far from the −150 mV mitochondrial membrane potential, which would tend to drive the membrane potential towards that equilibrium potential for anions, which we are presenting as somewhere around zero. Normally the mitochondrion has controlled conductance, but not if you had a selective channel that had an equilibrium potential far from where the resting potential is.

Nicholls: That happens at the plasma membrane where there is an 'infinite' pool of ions on either side. If you open up a channel, the membrane potential gets clamped. Instead you are focusing on the movement of the superoxide anion which is present in tiny amounts.

O'Rourke: There are many anions present that can get through this channel. Even small metabolites such as malate have been shown to go through this inner membrane anion channel. The superoxide is going along with the flow. It is a very small concentration. This flow of anions contains a certain amount of superoxide that can be exported across the membrane. In our model we have some data to suggest that the superoxide is produced on the matrix side. First, our sensor is there, and this is getting oxidized in the matrix. We think there are oxidants produced in the matrix. Second, if we inhibit the benzodiazepine receptor we get a bigger increase in ROS in the flashed region, but we don't get any propagation outside of this. This is why we hypothesize that it is generated on the matrix face and not on the outside face of the inner membrane. The other theoretical problem with it being produced on the outside is that it is less likely to oscillate because the activator site is on the same side as the generation of the activator.

Nicholls: I still have problems understanding that there is enough ion movement to significantly depolarize the mitochondria.

Bernardi: This is the same point I made a couple of years ago. First of all, have you identified the charge-carrying anion, because without that you don't have a mechanism? I don't think there is more malate or glutamate inside mitochondria than there is outside.

O'Rourke: There's plenty of chloride. There's 20 mM outside and other anions inside.

CARDIAC FUNCTION 153

Bernardi: As long as mitochondria are energized they will exclude chloride. If you open an ion channel that can drive chloride uptake it would hyperpolarize, not depolarize.

O'Rourke: The model has an outward rectifying chloride channel based on the patch clamp studies of Borecky and Siemen (Borecky et al 1997). You don't need many ions to move to change the membrane potential. In the sarcolemmal membrane, for example, there are very few cations moving to depolarise the membrane potential even though there is a large K^+ conductance there at all times. It depends on the impedance of the membrane.

Nicholls: When I started trying to understand chemiosmosis and mitochondrial membrane potentials, I was enormously confused about what happens at the plasma membrane where the classic Hodgkin–Huxley equation holds, with infinite compartments and where the flux of a few ions can depolarise, and what happens in the mitochondrion with an 'infinitessimally' small matrix compartment.

O'Rourke: It's true that there are constraints on volume and ion movement.

Brand: I agree with David Nicholls and Paolo Bernardi: there is a fundamental difference between the plasma membrane and the way that it responds to an anion channel opening, and the mitochondrial membrane. You have to move large amounts of ion across the membrane if it is the mitochondrial membrane, and because of the small matrix volume, they are just not there. I don't buy into the idea that increasing the Cl^- conductance would change the potential very much because this potential is dominated by the electrogenic electron transport chain, not by secondary ion movements. It is quite different from the plasma membrane, where the potential is dominated by secondary ion movements, not by the primary electrogenic sodium pump activity.

O'Rourke: Would you agree that the PTP can change the mitochondrial membrane potential? I should point out that in the swelling assay studies this was a promiscuous channel. It was not very selective even for anions and cations. It was a 4:1 cation:anion selectivity. I think many ions can move through this channel when it is open. I come from an ion channel background so I am glad to hear these criticisms.

Halestrap: There is one way you can account for it: if phosphate goes in with a proton on the phosphate transporter and comes out on an anion channel you can effectively get a net proton movement. I don't buy this story, but I can see that it has some merit. For example, we do know that modifying single thiol groups on mitochondrial substrate transporters can turn them into channels (Dierks et al 1990). We also know that ROS can modify thiol groups. So one can conceive that a substrate transporter could become an anion channel under conditions of oxidative stress.

O'Rourke: We are not making any judgement of which particular protein constitutes IMAC because we don't know the structure.

Rich: What about hydroxide as a candidate anion that goes through the channel as it is there in an unlimited supply?

Lemasters: Photodamage-induced injury is not the same as reperfusion injury. Photodamage is primarily mediated by singlet oxygen.

O'Rourke: The only photodamage is in those 50 mitochondria in the cell.

Lemasters: That is the initiator. We have looked at ischaemia–reperfusion in myocytes, and there are some interesting parallels. During ischaemia, Ca^{2+} goes up in the cytosol and the mitochondria, saturating the calcium indicators. During reperfusion, both cytosolic and mitochondrial Ca^{2+} recover quickly, and the mitochondria repolarize. The mitochondria remain relatively stable, except that they are generating free radicals at an increased and relatively steady rate. Then after 20–40 minutes, everything goes bad: the mitochondria depolarize, Ca^{2+} in the cytosol and mitochondria again increases to saturate the Ca^{2+} indicators. Somewhat later, the myocytes die. In these experiments, if we add permeability transition inhibitors, such as cyclosporin A or NIM811, we prevent depolarization and cell death. We also prevent the Ca^{2+} dysregulation, but we do not prevent the production of ROS. If we use antioxidants to block the ROS, we still prevent the depolarization, Ca^{2+} dysregulation and cell death. In this model, there is a steady production of ROS after reperfusion that reaches a threshold to induce the permeability transition. Flickering of the depolarization indicative of transient permeability transition pore opening precedes the sustained depolarization (Kim et al 2006).

O'Rourke: I view it as being a hierarchy of ROS-induced targets. The IMAC is even activated during ischaemia, because with 4-chloro-diazepam you can delay the depolarization of mitochondrial membrane potential in a whole heart, and get a better recovery on reperfusion, whereas cyclosporin doesn't do that. It only helps on reperfusion. In the first few minutes of reperfusion we are getting a mechanism that is not really PTP dependent, and then the PTP opens after a short delay. I hope we will get to a point where not every mitochondrial depolarization is attributed to the PTP.

Jacobs: Can you explain to me what is involved in propagating a signal from one mitochondria to the next, which leads to the lights going out?

O'Rourke: It is ROS diffusing from one mitochondrion to its neighbours, based on this percolation theory. This is the macroscopic cue determining which mitochondria are going to depolarize.

Jacobs: In this case, ROS is doing the same thing from the outside to the neighbour mitochondrion as it is doing from the inside to the one that is propagating the signal.

O'Rourke: This is why we have the activator site on the channel depicted on the outside in our model. We haven't addressed the outer membrane permeability, and whether a change is needed in this. We are presuming that the superoxide can get through the outer membrane as well.

Nicholls: So you are saying that mitochondria full of superoxide are waiting until they see external superoxide. They release this and a chain reaction proceeds.

O'Rourke: It is a regenerative wave in the system. As long at those mitochondria are at this threshold. I wouldn't say that it is just because superoxide is sitting there: it is also a depletion of the scavenger pool. We can get to this critical state in many ways, such as substrate depletion or reduced glutathione depletion. If cells are treated with Diamide, there is a certain critical level of glutathione that is reached, and then these oscillations occur, as well as eventual depolarization. It is the balance between the ROS production and the amount of scavenger that is present. If you load up the cytoplasm with TMPYP, or get the reduced glutathione levels high, this propagation won't occur.

Duchen: For the changes in mitochondrial potential to be translated into the dysrhythmia seen at the plasma membrane, you would need very fast turnover of ATP to regulate the plasmalemmal K-ATP channel.

O'Rourke: It isn't changes in ATP that matter as much as changes in the ATP:ADP ratio. We think it is mainly the increase in ADP that causes the change in sensitivity of the K-ATP channel. Mitochondrial uncoupling can rapidly activate sarcolemmal K-ATP channels, even if there is 5 mM ATP in the pipette solution. We have recently modelled this in an integrated model in which we put the mitochondria into a model of electrophysiology and Ca^{2+} handling. We can reproduce the effects on the action potential just by the burst of ADP that is produced by the mitochondria as they reverse and begin to consume ATP.

Adam-Vizi: I'd like to comment on ROS generation. People usually show schemes pointing to two major sites of ROS generation: the respiratory chain and complex III. This is true if one refers to data obtained with isolated mitochondria, where complex III is blocked with antimycin and the result is huge ROS generation. But if you are discussing physiologically relevant ROS generation, this is narrowed to complex I. For complex III to produce ROS it must be blocked almost totally, which is unrealistic *in vivo*.

O'Rourke: I disagree. I think the physiological cases show ROS generation from complex III but not complex I. We have never seen any state where if we reduce complex I by adding rotenone, this results in increased ROS production. We never get the production from complex I that can be obtained in an isolated mitochondrial system. This requires a very high level of reduction of complex I. Kushnareva had a paper (Kushnareva et al 2002) that suggested that it was even more reduced than NADH. This level of reduction doesn't usually occur in heart cells. For complex III, there is a paper by Turrens & Lehninger (1985) in which they took isolated mitochondria and depleted them of cytochrome c, and then added it back. They had a nice titration of VO2 as they did this. There was a linear correlation between ROS production in complex III and VO2. We believe this model. It is controversial.

Adam-Vizi: You can easily titrate complex III activity and measure ROS production. It can be blocked up to 70% and there is no ROS production.

O'Rourke: In our experiments, blocking the entry of electrons into complex III with myxothiazol or reducing the downstream electron acceptors, e.g. with cyanide, suppresses mitochondrial ROS production and prevents mitochondrial depolarization, consistent with superoxide generation at complex III.

References

Borecky J, Jezek P, Siemen D 1997 108-pS channel in brown fat mitochondria might be identical to the inner membrane anion channel. J Biol Chem 272:19282–19289

Dierks T, Salentin A, Kramer R 1990 Pore-like and carrier-like properties of the mitochondrial aspartate/glutamate carrier after modification by SH-reagents—evidence for a preformed channel as a structural requirement of carrier-mediated transport. Biochim Biophys Acta 1028:281–288

Kim JS, Jin Y, Lemasters JJ 2006 Reactive oxygen species, but not Ca2+ overloading, trigger pH- and mitochondrial permeability transition-dependent death of adult rat myocytes after ischemia-reperfusion. Am J Physiol Heart Circ Physiol 290:H2024–H2034

Kushnareva Y, Murphy AN, Andreyev A 2002 Complex I-mediated reactive oxygen species generation: modulation by cytochrome c and NAD(P)+ oxidation-reduction state. Biochem J 368:545–553

Turrens JF, Alexandre A, Lehninger AL 1985 Ubisemiquinone is the electron donor for superoxide formation by complex III of heart mitochondria. Arch Biochem Biophys 237:408–414

The mitochondrial permeability transition pore

Paolo Bernardi and Michael Forte*

*Department of Biomedical Sciences and CNR Institute of Neurosciences, University of Padova, Viale Giuseppe Colombo 3, I-35121 Padova, Italy and *Vollum Institute, Oregon Health and Science University, Portland, OR 97239, USA*

Abstract. The mitochondrial permeability transition pore (PTP) is a high conductance channel of the inner membrane whose opening leads to an increase of permeability to solutes with molecular masses up to about 1500 Da, the 'permeability transition'. This potentially catastrophic event has long been known, yet the molecular bases for its occurrence remain unsolved despite its established importance in several *in vivo* models of pathology. Recent studies based on inactivation of genes encoding putative pore components (such as the adenine nucleotide translocators and the voltage-dependent anion channel) have raised major questions about the involvement of these proteins in PTP formation, yet they have conclusively demonstrated the role of matrix cyclophilin D as the mitochondrial receptor for the desensitizing effects of cyclosporin A. While the nature of the components forming the PTP remains controversial, the identification of novel inhibitors that can be used as affinity labels is offering new perspectives towards the molecular definition of the PTP.

2007 Mitochondrial biology: new perspectives. Wiley, Chichester (Novartis Foundation Symposium 287) p 157–169

The mitochondrial permeability transition (PT) is an increase of mitochondrial inner membrane permeability to solutes with molecular masses up to about 1500 Da. If long-lasting, this event is followed by mitochondrial depolarization, matrix swelling, depletion of pyridine nucleotides, outer membrane rupture and release of intermembrane proteins (Bernardi 1999, Bernardi et al 2006, Crompton 1999, Halestrap et al 2004). The molecular structure of the pore remains undefined, as should become clear from the discussion of the most studied candidate proteins, i.e. the adenine nucleotide translocator (ANT), the voltage-dependent anion channel (VDAC) and the peripheral benzodiazepine receptor (PBR), while the role of cyclophilin D (CyP-D) as a modulator cannot be questioned as shown by genetic inactivation of the *Ppif* gene encoding for CyP-D in the mouse (Baines et al 2005, Basso et al 2005, Nakagawa et al 2005, Schinzel et al 2005).

Adenine nucleotide translocator

The permeability transition pore (PTP) is modulated by ligands of the ANT. Atractylate inhibits the ANT and favours PTP opening while bongkrekate (which also inhibits the ANT) instead favours PTP closure. These findings led to the suggestion that the PTP may be directly formed by the ANT (Halestrap & Brenner 2003), a hypothesis that found some support in partial purifications of the ANT followed by reconstitution of its pore-forming activity in a variety of assays. Although intriguing, we think that the relevance of these observations remains unclear because the purification was only partial, and it is difficult to exclude that the PTP-like activity was due to other species less represented than the abundant ANT (reviewed in Bernardi et al 2006).

A more straightforward approach testing the involvement of the ANT in PTP formation was based on genetic inactivation of genes encoding ANT isoforms. A study of mouse mitochondria lacking each of the two ANT isoforms present in this species revealed that a Ca^{2+}-dependent PT took place, and that the ANT-null mitochondria underwent the same swelling response as those from wild-type animals. However, ANT-deficient mitochondria required a higher Ca^{2+} load to initiate PTP-dependent mitochondrial permeabilization when compared to wild-type mitochondria (Kokoszka et al 2004). The PT of ANT-null mitochondria was fully inhibitable by CsA and could be triggered by H_2O_2 and diamide, indicating that the ANT is neither the obligatory binding partner of CyP-D nor the site of action of oxidants (Kokoszka et al 2004).

It has been argued that a low, undetectable level of ANT expression could have allowed the PTP in 'ANT-null' mitochondria (Halestrap 2004). This hypothesis overlooks the experimental findings that the PTP became *insensitive* to opening by atractylate and to closure by ADP. It is very difficult for us to envisage how these hypothetical ANT molecules, that would remain in mitochondria after inactivation of the two isoforms targeted in the studies outlined above, would not respond to atractylate and ADP, and yet be able to promote a CsA-sensitive PT. In other words, the lack of response of the PTP to atractylate and bongkrekate is the best internal control for the absence of ANT molecules in the mutant mitochondria. A recent observation in the anoxia-tolerant brine shrimp *Artemia franciscana* is also relevant to this discussion. Despite a rather remarkable Ca^{2+} uptake capacity and the presence of ANT, VDAC and CyP-D, mitochondria isolated from this organism do not undergo a PT, suggesting that the PTP may be composed of proteins (or factors) that remain to be identified (Menze et al 2005).

We believe that the experiments in ANT-null mitochondria (Kokoszka et al 2004) demonstrate beyond reasonable doubt that the ANT is not an essential component of the PTP; and that these results cannot be explained by the existence of 'latent' ANT molecules that would possess the unprecedented ability of not binding ADP, atractylate and bongkrekate. If the ANT is not a mandatory component of the PTP, why do the latter compounds affect the pore? We think that the changes of the surface potential following the transition of the abundant translocase between the 'm' conformation (ADP or bongkrekate-bound) and the 'c' conformation (atractylate-bound) (Schultheiss & Klingenberg 1984) may explain pore modulation by ANT ligands within the framework of the pore voltage-dependence (Bernardi 1999).

Voltage-dependent anion carrier

The outer membrane may be required for the PT, since pore opening induced by sulfhydryl reagents is not observed in mitoplasts (Le Quôc & Le Quôc 1982). The outer membrane element involved in PTP formation has been suggested to be VDAC based on the following (see Bernardi et al 2006 for the original literature): (i) the electrophysiological properties of purified VDAC are similar to those of the PTP, which also shares with VDAC an estimated pore diameter of 2.5–3.0 nm; (ii) like the PTP, VDAC is modulated by NADH, Ca^{2+}, glutamate and hexokinase; (iii) chromatography of mitochondrial extracts on a CyP-D affinity matrix allowed purification of VDAC and the ANT, which in the presence of CyP-D catalysed CsA-sensitive permeabilization of liposomes to solutes; (iv) Ro 68-3400, a high-affinity inhibitor of the PTP identified through the screening of a chemical library, labelled a protein of approximately 32 kDa that was identified as VDAC1.

In order to assess the role of VDAC1 in PTP formation and activity, we have studied the properties of mitochondria from $Vdac1^{-/-}$ mice. The basic properties of the PTP were indistinguishable from those displayed by mitochondria from wild-type mice, including inhibition by Ro 68–3400 which labelled identical proteins of 32 kDa in wild-type and $Vdac1^{-/-}$ mitochondria (Krauskopf et al 2006). Since the labelled protein could be separated from all VDAC isoforms, the putative 32 kDa component of the PTP cannot be VDAC. While these results do not rule out that VDAC is part of the PTP, especially given the fact that all eukaryotic genomes encode multiple VDAC isoforms, they do indicate that VDACs are not the targets for PTP inhibition by Ro 68–3400 (Krauskopf et al 2006).

Peripheral benzodiazepine receptor

The peripheral benzodiazepine receptor (PBR) is an 18 kDa hydrophobic protein located in the outer mitochondrial membrane that was initially identified as a binding site for benzodiazepines in tissues that lack GABA receptors. Involvement of the PBR in PTP function was suggested by the close association of this protein with the putative pore components VDAC and ANT (McEnery et al 1992). However, binding of high affinity ligands to the PBR does not require either of these proteins (Lacapere et al 2001). Evidence that the PBR may play a role in PTP formation or regulation was obtained in patch clamp analysis of mitoplasts, which are largely devoid of outer membranes. Treatment with PBR ligands affected the single channel activity of the PTP in mitoplasts (Kinnally et al 1993) but it is difficult to see how an inner membrane channel could be affected by the outer membrane PBR receptor since patched membranes would rarely contain outer membrane fragments. It is therefore reasonable to wonder if the PBR ligands investigated might directly affect the PTP and not require association with the PBR.

Indeed, the effects of these drugs in promoting opening of the PTP (and apoptosis) depends on the cell type examined and the concentrations of PBR ligand tested (Berson et al 2001, Bono et al 1999, Chelli et al 2001, Hirsch et al 1998, Pastorino et al 1994, Szabó & Zoratti 1993). The same compounds may have a PTP-inhibiting and/or antiapoptotic effect in other models (Leducq et al 2003, Parker et al 2002). Further, individual PBR ligands can have different effects on the same cell. For example, Ro5-4864 and PK11195 can either stimulate or inhibit apoptosis, respectively, with addition of Ro5-4864 overcoming the antiapoptotic effect of PK11195 (Bono et al 1999).

Recent results suggesting that PBR ligands can generate cellular phenotypes that are independent of the PBR have muddled any definitive notions as to the physiological role of the PBR and its involvement in PTP activity based on the use of these 'specific' ligands (Gonzalez-Polo et al 2005, Hans et al 2005, Kletsas et al 2004). In summary, it cannot be ruled out that the PBR is part of the PTP but conclusions based on the exclusive use of PBR ligands should be viewed and interpreted with caution.

Cyclophilin D

Opening of the PTP is inhibited by CsA after binding to CyP-D, a matrix peptidyl-prolyl *cis-trans* isomerase. CyP-D may be involved in modulation of the PTP affinity for Ca^{2+} since higher concentrations of CsA are required to inhibit spreading of the PTP to a population of mitochondria when the Ca^{2+} load is increased (Bernardi et al 1992). Available evidence suggests that calcineurin is not involved in the effects of CsA on the PTP because CsA derivatives have been described that bind CyP-D and desensitize the pore, but do not inhibit calcineurin (Bernardi et al 2006). However, calcineurin can also affect mitochondrial function through the dephosphorylation of BAD and the release of apoptogenic proteins (Wang et al 1999).

Conclusive results on the role of CyP-D in regulation of the PTP have been obtained after inactivation of the *Ppif* gene, which encodes CyP-D in the mouse (Baines et al 2005, Basso et al 2005, Nakagawa et al 2005, Schinzel et al 2005). In all studies, ablation of CyP-D increased the Ca^{2+} retention capacity, i.e. the threshold Ca^{2+} load required to open the PTP, which became identical to that of CsA-treated wild type mitochondria, while no effect of CsA was observed in $Ppif^{-/-}$ mitochondria (Baines et al 2005, Basso et al 2005, Nakagawa et al 2005, Schinzel et al 2005). Provided that a permissive load of Ca^{2+} (which is larger for the $Ppif^{-/-}$ mitochondria) had been accumulated, the sensitivity to other regulatory factors was not changed appreciably, while the sensitivity to oxidative stress was enhanced (Basso et al 2005). These findings demonstrate that CyP-D is a regulator but not a core component of the PTP, whose structure is unlikely to be altered by the absence of CyP-D. Based on these results the effect of CsA is best described as 'desensitization' rather than inhibition of the PTP, since its effects (similar to the lack of CyP-D) can be overcome by increasing the Ca^{2+} load. At variance with the conclusions of influential reviews (Green 2005, Halestrap 2005), the *in vivo* studies on $Ppif^{-/-}$ mice can only be interpreted in terms of the role of CyP-D, not of the PTP, in cell death. Indeed, all studies agree that the PTP can form and open in the absence of CyP-D, provided that a permissive Ca^{2+} load is accumulated (Baines et al 2005, Basso et al 2005, Nakagawa et al 2005, Schinzel et al 2005). Thus, the inference that PTP opening does not take place because CyP-D is absent has not been documented *in vivo*, an issue that questions the conclusion that the PTP participates in cell death pathways only in response to a restricted set of challenges.

What is the permeability transition pore?

We think that none of the candidate pore components has passed rigorous genetic testing. Irrespective of its molecular nature, however, a comment is in order on the idea that the pore forms at 'contact sites' between the inner and outer mitochondrial membranes and that it spans both membranes (Zamzami & Kroemer 2001). We believe that the idea that the PTP forms at contact sites is based on a set of assumptions rather than on established facts, and should be considered with caution since the very existence of points of fusion between the outer and inner membranes has been questioned (Mannella 2006). The second point does not take into account that the permeability pathway resulting from a pore spanning both membranes would directly connect the matrix with the cytosol allowing release of matrix solutes but not of cytochrome c and other intermembrane proapoptotic proteins, which is instead observed when the pore opens.

In our opinion, the PT is primarily an inner membrane event. This does not preclude the possibility that outer membrane components could regulate the activity of the PTP, possibly through protein–protein interactions that could depend on a specific conformation conferred by cytosolic regulator(s). As explained in more detail elsewhere, the outer membrane could confer regulatory features to the PTP without necessarily providing a permeability pathway for solute diffusion (Bernardi et al 2006).

Conclusions

We think that the molecular nature of the PTP has not been solved, and that the field may benefit from a fresh look at this old problem. On the other hand, based on impressive results in *in vivo* models of disease, the pathophysiological importance of the PTP is hard to question (reviewed in Bernardi et al 2006). We are confident that the gap between our structural and functional understanding of the PTP will be filled reasonably soon.

Acknowledgements

Research in our laboratories is supported by the Italian Ministry for the University (PB), AIRC Grant 1293 (PB), Telethon-Italy Grant GGP04113 (PB) and the National Institutes of Health—Public Health Service (USA) Grant GM69883 (PB and MF).

References

Baines CP, Kaiser RA, Purcell NH et al 2005 Loss of cyclophilin D reveals a critical role for mitochondrial permeability transition in cell death. Nature 434:658–662

Basso E, Fante L, Fowlkes J, Petronilli V, Forte MA, Bernardi P 2005 Properties of the permeability transition pore in mitochondria devoid of cyclophilin D. J Biol Chem 280:18558–18561

Bernardi P 1999 Mitochondrial transport of cations: channels, exchangers and permeability transition. Physiol Rev 79:1127–1155

Bernardi P, Vassanelli S, Veronese P, Colonna R, Szabo I, Zoratti M 1992 Modulation of the mitochondrial permeability transition pore. Effect of protons and divalent cations. J Biol Chem 267:2934–2939

Bernardi P, Krauskopf A, Basso E et al 2006 The mitochondrial permeability transition from in vitro artifact to disease target. FEBS J 273:2077–2099

Berson A, Descatoire V, Sutton A et al 2001 Toxicity of alpidem, a peripheral benzodiazepine receptor ligand, but not zolpidem, in rat hepatocytes: role of mitochondrial permeability transition and metabolic activation. J Pharmacol Exp Ther 299:793–800

Bono F, Lamarche I, Prabonnaud V, Le Fur G, Herbert JM 1999 Peripheral benzodiazepine receptor agonists exhibit potent antiapoptotic activities. Biochem Biophys Res Commun 265:457–461

Chelli B, Falleni A, Salvetti F, Gremigni V, Lucacchini A, Martini C 2001 Peripheral-type benzodiazepine receptor ligands: mitochondrial permeability transition induction in rat cardiac tissue. Biochem Pharmacol 61:695–705

Crompton M 1999 The mitochondrial permeability transition pore and its role in cell death. Biochem J 341:233–249

Gonzalez-Polo RA, Carvalho G, Braun T et al 2005 PK11195 potently sensitizes to apoptosis induction independently from the peripheral benzodiazepin receptor. Oncogene 24:7503–7513

Green DR 2005 Apoptotic pathways: ten minutes to dead. Cell 121:671–674

Halestrap AP 2004 Mitochondrial permeability: dual role for the ADP/ATP translocator? Nature 430, 1 p following 983

Halestrap A 2005 Biochemistry: a pore way to die. Nature 434:578–579

Halestrap AP, Brenner C 2003 The adenine nucleotide translocase: a central component of the mitochondrial permeability transition pore and key player in cell death. Curr Med Chem 10:1507–1525

Halestrap AP, Clarke SJ, Javadov SA 2004 Mitochondrial permeability transition pore opening during myocardial reperfusion—a target for cardioprotection. Cardiovasc Res 61:372–385

Hans G, Wislet-Gendebien S, Lallemend F et al 2005 Peripheral benzodiazepine receptor (PBR) ligand cytotoxicity unrelated to PBR expression. Biochem Pharmacol 69:819–830

Hirsch T, Decaudin D, Susin SA et al 1998 PK11195, a ligand of the mitochondrial benzodiazepine receptor, facilitates the induction of apoptosis and reverses Bcl-2-mediated cytoprotection. Exp Cell Res 241:426–434

Kinnally KW, Zorov DB, Antonenko YN, Snyder SH, McEnery MW, Tedeschi H 1993 Mitochondrial benzodiazepine receptor linked to inner membrane ion channels by nanomolar actions of ligands. Proc Natl Acad Sci USA 90:1374–1378

Kletsas D, Li W, Han Z, Papadopoulos V 2004 Peripheral-type benzodiazepine receptor (PBR) and PBR drug ligands in fibroblast and fibrosarcoma cell proliferation: role of ERK, c-Jun and ligand-activated PBR-independent pathways. Biochem Pharmacol 67:1927–1932

Kokoszka JE, Waymire KG, Levy SE et al 2004 The ADP/ATP translocator is not essential for the mitochondrial permeability transition pore. Nature 427:461–465

Krauskopf A, Eriksson O, Craigen WJ, Forte MA, Bernardi P 2006 Properties of the permeability transition in VDAC1$^{-/-}$ mitochondria. Biochim Biophys Acta 1757:590–595

Lacapere JJ, Delavoie F, Li H et al 2001 Structural and functional study of reconstituted peripheral benzodiazepine receptor. Biochem Biophys Res Commun 284:536–541

Le Quôc K, Le Quôc D 1982 Control of the mitochondrial inner membrane permeability by sulfhydryl groups. Arch Biochem Biophys 216:639–651

Leducq N, Bono F, Sulpice T et al 2003 Role of peripheral benzodiazepine receptors in mitochondrial, cellular, and cardiac damage induced by oxidative stress and ischemia-reperfusion. J Pharmacol Exp Ther 306:828–837

Mannella CA 2006 The relevance of mitochondrial membrane topology to mitochondrial function. Biochim Biophys Acta 1762:140–147

McEnery MW, Snowman AM, Trifiletti RR, Snyder SH 1992 Isolation of the mitochondrial benzodiazepine receptor: association with the voltage-dependent anion channel and the adenine nucleotide carrier. Proc Natl Acad Sci USA 89:3170–3174

Menze MA, Hutchinson K, Laborde SM, Hand SC 2005 Mitochondrial permeability transition in the crustacean Artemia franciscana: absence of a calcium-regulated pore in the face of profound calcium storage. Am J Physiol 289:R68–R76

Nakagawa T, Shimizu S, Watanabe T et al 2005 Cyclophilin D-dependent mitochondrial permeability transition regulates some necrotic but not apoptotic cell death. Nature 434:652–658

Parker MA, Bazan HE, Marcheselli V, Rodriguez de Turco EB, Bazan NG 2002 Platelet-activating factor induces permeability transition and cytochrome c release in isolated brain mitochondria. J Neurosci Res 69:39–50

Pastorino JG, Simbula G, Gilfor E, Hoek JB, Farber JL 1994 Protoporphyrin IX, an endogenous ligand of the peripheral benzodiazepine receptor, potentiates induction of the mitochondrial permeability transition and the killing of cultured hepatocytes by rotenone. J Biol Chem 269:31041–31046

Schinzel AC, Takeuchi O, Huang Z et al 2005 Cyclophilin D is a component of mitochondrial permeability transition and mediates neuronal cell death after focal cerebral ischemia. Proc Natl Acad Sci USA 102:12005–12010

Schultheiss HP, Klingenberg M 1984 Immunochemical characterization of the adenine nucleotide translocator. Organ specificity and conformation specificity. Eur J Biochem 143:599–605

Szabó I, Zoratti M 1993 The mitochondrial permeability transition pore may comprise VDAC molecules. I. Binary structure and voltage dependence of the pore. FEBS Lett 330:201–205

Wang H, Pathan N, Ethell IM et al 1999 Ca^{2+}-induced apoptosis through calcineurin dephosphorylation of BAD. Science 284:339–343

Zamzami N, Kroemer G 2001 The mitochondrion in apoptosis: how Pandora's box opens. Nat Rev Mol Cell Biol 2:67–71

DISCUSSION

Nicholls: What Peter Mitchell said many years ago is that we should be trying to disprove our hypotheses rather than prove them. You've done well on this count. You talked about what the PTP isn't. So what do you feel it is?

Bernardi: I am not going to tell you until I am sure! I don't know.

Martinou: I don't understand why you say that, because cyclosporin A works, then the PTP is causing the myopathy. Can you exclude that CyP-D has an action, independent of PTP regulation?

Bernardi: I am saying that inappropriate pore opening through CyP-D is clearly part of the pathogenesis. In this particular disease we have been very lucky. Something else is obviously going on, yet with cyclosporin we recover strength in the animal, and there is no question that in this disease the pore is part of the pathogenetic chain of events. Clearly there is also a feedback loop on the SR, but in this case CyP-D is the key.

Martinou: I wouldn't say that this means that the PTP is involved in this case.

Bernardi: I would be surprised if the pore is not the key, because we have recovered the ultrastructure by either treating myopathic animals with cyclosporin A or by crossing them with CyP-D null animals. We can show that in the fibres the mitochondria don't depolarize any more. I would have a hard time finding a better explanation.

Lemasters: How is it that the collagen deficiency causes the permeability transition?

Bernardi: That's the crucial question. We are investigating several pathways that are known to lead from integrins to function/dysfunction. One possibility is that the lack of collagen VI, which is a signalling molecule, will engage the mitochondrial pathway through Bcl-2 and Bax levels. We know that Bcl-2 goes down dramatically in the knockout fibres. This is a potential mechanism for sensitising the mitochondria to Ca^{2+} and oxidative stress. We know that caspases are activated downstream of the mitochondria. We know that cytochrome c is released. It looks like a push forward towards the apoptotic phenotype.

Lemasters: Why do the cells not become Ca^{2+} overloaded?

Bernardi: There is no increase of resting Ca^{2+}, but we can easily induce Ca^{2+} deregulation with oligomycin. We suspect that the stores are filled with Ca^{2+}. Maybe it is the SR, but for sure we have a back-up system with mitochondria that fails in this case. I am not saying that this is the cause of the disease. The cause of the disease is lack of collagen VI. I am saying that you can cure a disease by blocking an effector pathway that is downstream of the genetic lesion.

Nicholls: You are getting these effects with cyclosporin A. Are you getting the same rescue with specific permeability pore transition inhibitors that don't affect the immune response?

Bernardi: We haven't done the *in vivo* work yet, but in the cells all that I have shown with cyclosporin can be done with a ligand of cyclophilin that doesn't inhibit calcineurin. Calcineurin is therefore not necessarily part of the picture so far.

Youle: When you rescue the disease with a double knockout, do you prevent cytochrome c release?

Bernardi: I haven't done this yet, but I think we would. This would need to be done in myoblast cultures. Fibres are a big problem.

Beal: You said that there is a tendency for apoptosis in the cyclophilin cells. My reading of those knockout papers is that PTP seemed to be critical in necrotic cell death, and didn't play a role in apoptosis in these fibroblast knockouts.

Bernardi: This is an example of an oversimplification. We have *in vivo* models of TNFα-dependent hepatitis, and find some cells are necrotic while others are apoptotic. Pierluigi Nicotera showed that depending on the ATP levels you can modulate the rheostat for necrosis/apoptosis (Ankarcrona et al 1994) and I wouldn't take experiments done on isolated cells at face value. Those data on Bax and Bak are done with H_2O_2, and this is a major problem. The levels of expression of catalase in these cell lines is so variable. Our knockouts have more catalase than the controls, so they are protected from H_2O_2 damage not because the pore is less sensitive but rather because they scavenge the H_2O_2 much better, an issue that has not been addressed in the *Nature* papers you are referring to.

Beal: So you think that cyclophilin can play a role in PTP-mediated apoptosis?

Bernardi: The pore can play a role even in apoptosis.

Beal: So you can't use this as a method for trying to distinguish those modes of cell death. I am interested in this because we'd like to cross these mice into the ALS or the Huntington's mice. The answer probably won't be interpretable from what you are saying.

Bernardi: Not in terms of the mechanism. The consequences of pore opening very much depend on the open time. With short open times you can have no release of cytochrome c but you can have ATP depletion and then the necrotic pathway. Longer open times can lead to intermediate states. It is not an all-or-nothing event.

Halestrap: Clearly, for the normal programmed cell death that occurs in development, the CyP-D knockout mouse behaves normally. When it comes to ischaemic stress, a mild stress may lead to transient pore opening and with enough cytochrome c release there will be apoptosis because you can maintain cellular ATP levels. Once you get past a certain point of pore opening there are more mitochondria hydrolysing ATP than producing it and the result is necrosis, even if there is caspase activation. I agree with about 80% of what Paolo has said! I agree that CyP-D is critical, but it is only sensitising pore opening. I don't think VDAC is essential but it may be regulatory. The area we disagree on is the role of the ANT. I will be the first to admit that it is not the sole membrane component. However, the data in the ANT knockout mitochondria show that they are far less sensitive to pore opening under some conditions. The pore is no longer blocked by adenine nucleotides. This is a fundamental control mechanism. If you deplete adenine nucleotides in normal mitochondria, the pore goes into hyperdrive. So in a normal mitochondrion ANT is playing a critical role. I should add that the antibody we used to try to prove that the ANT was the main protein binding to cyclophilin also binds to another 32 kDa protein. We know from phenylarsine oxide columns that ANT and another 32 kDa membrane protein will bind to the column, the latter also being a member of the membrane transporter family. I suspect that either several members of the transporter family can form the pore, or there are hybrids, perhaps even where ANT and another one come together. So there is another protein in there, and it is presumably the one your Roche compound is binding to. I'd also say that ubiquinone binds to the ANT at the same concentrations as it inhibits the pore.

Nicholls: External adenine nucleotides certainly seem to protect against the PTP. There is a yin and yang of carboxyatractylate and bongkrekic acid which fits nicely with cytoplasmic nucleotide depletion. This seems to be incontrovertible. So how does one extrapolate this to a situation where there is no ANT?

Bernardi: I have a comment on the effect of ADP and bongkrekic acid on the one hand, and atractylate on the other. These are unnatural states of the ANT, because you are synchronizing the protein that is around 13% of the IMM mass in one specific state. Hagai Rottenberg showed many years ago (Rottenberg & Marbach 1990) that when atractylate is used to synchronize the ANT in the 'c' conformation, it neutralized surface charges, corresponding with depolarization. On the other hand, bongkrekic acid or ADP on the matrix side synchronize the ANT in the 'm' conformation corresponding with hyperpolarization. My interpretation is that the ANT conformation modifies the surface potential, affecting in turn the pore open probability because of its voltage-dependence.

Halestrap: There is a further point that this can't explain. If you look at the ADP sensitivity of the pore in the shrinking assay, where you have open matrix, phenylarsine oxide and oxidative stress change the ability of ADP to inhibit the pore. This process is definitely ADP binding to something. Then you look at the sensitivity to different adenine nucleotides and it matches that of the translocase.

Bernardi: It is an endless story. The same thing is seen for cyclosporin effects. I can't make any conclusions based on isolated mitochondria or pharmacology, particularly with reagents that are not so selective.

Larsson: I have a comment about the ANT knockout mouse. As far as I remember there was a double knockout. One was germline and the other was conditional. How do we know that there isn't a certain percentage of the protein left in this knockout? In tissues that can proliferate we have seen that under some circumstances you can select cell clones that are not recombined. Could this be a potential problem with the knockout studies?

Bernardi: These mitochondria cannot undergo state IV–state III transitions with added ADP. For the very reason that the pore is completely insensitive to atractylate and ADP, I can't see how those molecules could be responsible for the transition pore in these mitochondria and yet not be responsive to atractylate.

Halestrap: I would agree with that, but they can't be totally ANT knockout in terms of their physiology. They have ANT4 in them, so functionally they have ANT (Da Cruz et al 2003). However, this doesn't affect your argument.

Nicholls: The question is, is the mouse really what you think it is? If it is a surviving mouse, how is it surviving?

Bernardi: I think they have the ATP magnesium phosphate co-transporter.

Halestrap: They also have ANT4. It was found in the proteome of mouse liver mitochondria (Da Cruz et al 2003).

Bernardi: Is that ANT insensitive to atractylate?

Halestrap: Yes.

Bernardi: Then it's not an explanation.

Nicholls: It is a major problem. You have a mouse that allegedly has no adenine nucleotide translocator and therefore cannot couple its mitochondria to supplying ATP to the cell. Yet it appears to be alive.

Schon: Is there a PTP in yeast?

Bernardi: No. There is something similar in some strains, but it is cyclosporin insensitive. It is not Ca^{2+} dependent and we can't study it as a *bona fide* model of the PTP.

Lemasters: One of the remarkable things, which none of the models explains very well, is that you have such a wide and diverse range of reactive chemicals that induce the permeability transition or sensitize to it. Also, pore-forming peptides can induce a permeability transition-like activity in the mitochondria (Pfeiffer et al 1995, He & Lemasters 2002). Vercesi has emphasized that protein damage is involved in PTP formation (Fagian et al 1990). One idea we have proposed is that the damage to integral membrane proteins leads to misfolding and exposure of hydrophilic surfaces in the bilayer that then aggregate (He & Lemasters 2002). This aggregation creates an aqueous pore, but a chaperone system is present that blocks pore conductance. I'm glad to hear Paolo call CyP-D a chaperone because it is a *cis–trans* peptidyl proline isomerase, namely a foldase. Cyclophilin binding to the nascent pores formed from misfolded membrane proteins would presumably confer to the pore complex its various properties, particularly sensitivity to Ca^{2+}. Adding Ca^{2+} to this closed pore complex causes pore opening, and then you would have the permeability transition. This event is mediated by CyP-D. By inference, if damaged proteins continue to form nascent pores, at some point you have more nascent pores than chaperones to regulate them. In this case you would get swelling that is independent of Ca^{2+} and insensitive to cyclosporin A. There are examples of this in the literature (Pfeiffer et al 1995, He & Lemasters 2002, Gadd et al 2006, Kowaltowski et al 1997). With low inducers you get cyclosporin A sensitivity and swelling that requires Ca^{2+}. At the higher amount of the same inducers you just get swelling, and Ca^{2+} chelation or cyclosporin A doesn't do anything. The same is true if you add endogenous pore-forming peptides, such as mastoparan or alamethicin. Since the ANT is the most abundant protein in the inner membrane and may be vulnerable to oxidative stress, damage and misfolding, the translocator is likely often involved in pore formation. If you take the translocator away, there are other integral membrane proteins that can get damaged, perhaps requiring somewhat more stress but giving rise to fundamentally similar phenomena. Hence you get the pore with the ANT knockout and you also get swelling with the CyP-D knockout. In the misfolded protein model, CyP-D acts as the Ca^{2+} sensor. If the sensor is removed, then more Ca^{2+} induction is needed to see the same thing, which is consistent with Paolo's results.

Bernardi: The pore can be studied by electrophysiology at the single channel level. It has a well characterized and reproducible behaviour. I have a problem believing that you can generate a well behaved pore with any disrupted membrane protein. It may be an aesthetic problem: my aesthetics tell me it is a well behaved pore and we will find it!

References

Ankarcrona M, Dypbukt JM, Bonfoco E et al 1995 Glutamate-induced neuronal death: a succession of necrosis or apoptosis depending on mitochondrial function. Neuron 15:961–973

Da Cruz S, Xenarios I, Langridge J, Vilbois F, Parone PA, Martinou JC 2003 Proteomic analysis of the mouse liver mitochondrial inner membrane. J Biol Chem 278:41566–41571

Fagian MM, Pereira-da-Silva L, Martins IS, Vercesi AE 1990 Membrane protein thiol cross-linking associated with the permeabilization of the inner mitochondrial membrane by Ca2+ plus prooxidants. J Biol Chem 265:19955–19960

Gadd ME, Broekemeier KM, Crouser ED, Kumar J, Graff G, Pfeiffer DR 2006 Mitochondrial iPLA2 activity modulates the release of cytochrome c from mitochondria and influences the permeability transition, J Biol Chem 281:6931–6939

He L, Lemasters JJ 2002 Regulated and unregulated mitochondrial permeability transition pores: a new paradigm of pore structure and function? FEBS Lett 512:1–7

Kowaltowski AJ, Vercesi AE, Castilho RF 1997 Mitochondrial membrane protein thiol reactivity with N-ethylmaleimide or mersalyl is modified by Ca2+: correlation with mitochondrial permeability transition. Biochim Biophys Acta 1318:395–402

Pfeiffer DR, Gudz TI, Novgorodov SA, Erdahl WL 1995 The peptide mastoparan is a potent facilitator of the mitochondrial permeability transition. J Biol Chem 270:4923–4932

Rottenberg H, Marbach M 1990 Regulation of Ca2+ transport in brain mitochondria. I. The mechanism of spermine enhancement of Ca2+ uptake and retention. Biochim Biophys Acta 1016:77–86

Mechanisms of mitochondrial outer membrane permeabilization

Dominic James, Philippe A. Parone, Olivier Terradillos, Safa Lucken-Ardjomande, Sylvie Montessuit and Jean-Claude Martinou

Department of Cell Biology, University of Geneva, Quai Ernest-Ansermet 30, Geneva, Switzerland

Abstract. In response to many apoptotic stimuli, Bcl-2 family pro-apoptotic members, such as Bax and Bak, are activated. This results in their oligomerization, permeabilization of the outer mitochondrial membrane, and release of many proteins that are normally confined in the mitochondrial inter-membrane space. Among these proteins are cytochrome c, Smac/DIABLO, OMI/HtrA2, AIF and endonuclease G. Mitochondrial outer membrane permeabilization (MOMP) is also associated with fragmentation of the mitochondrial network. The mechanisms that lead to the oligomerization of pro-apoptotic members of the Bcl-2 family and to MOMP are still unclear and the role of mitochondrial fission in these events remains elusive.

2007 Mitochondrial biology: new perspectives. Wiley, Chichester (Novartis Foundation Symposium 287) p 170–182

Mitochondria play a critical role in the regulation of apoptosis through the release of apoptogenic proteins such as cytochrome c, Smac/DIABLO, HtrA2/Omi, endonuclease G and AIF (Green & Reed 1998, Danial & Korsmeyer 2004, Ekert & Vaux 2005). The release of these proteins during apoptosis is regulated by a subclass of Bcl-2 proteins (Vander Heiden & Thompson 1999, Cory & Adams 2000), including Bax and Bak. These proteins seem to be in an inactive state in healthy cells, with Bax predominantly found in the cytosol. However, during apoptosis induced by various death stimuli including DNA damage or trophic factor deprivation, they are activated by a process requiring BH3-only Bcl-2 family members. It is thought that BH3-only proteins either bind and sequester Bcl-2 anti-apoptotic proteins (this is the case for Bad and Puma), or bind to and directly activate pro-apoptotic proteins (tBid for example) (Huang & Strasser 2000, Letai et al 2002, Terradillos et al 2002). This results in the inactivation of Bcl-2 anti-apoptotic proteins and in the oligomerization of Bax and Bak in the mitochondrial outer membrane (MOM) with a concomitant release of apoptogenic factors from the mitochondria (Desagher et al 1999, Eskes et al 2000, Kuwana et al 2002).

How permeabilization of the MOM (MOMP) occurs during apoptosis remains a matter of debate and has been extensively studied (Martinou & Green 2001). Recently, a new model has emerged based on the discovery that mitochondria fragment during cell death (Martinou et al 1999, Frank et al 2001, Jagasia et al 2005). According to this model, the fission of mitochondria would be necessary for MOMP (Youle & Karbowski 2005). Nevertheless, it is still not clear whether the fragmentation of mitochondria precedes or follows the release of apoptogenic factors (Arnoult et al 2005).

Mitochondrial fission and fusion are normal and frequent events in healthy cells. The protein machinery that underlies mitochondrial fission has been well characterised and extensively reviewed (Rube & van der Bliek 2004). In mammalian cells, at least three proteins, Drp1, hFis1 and MTP18 are required for this process. The dynamin-related protein Drp1 is a large cytosolic GTPase that translocates to the mitochondria where it couples GTP hydrolysis with scission of the mitochondrial tubule (Smirnova et al 1998, Pitts et al 1999). It is still unclear whether the process of mitochondrial fission, which systematically occurs in Bax/Bak-dependent apoptosis, is required for cell death (Bossy-Wetzel et al 2003, Martinou & Youle 2006). In this communication, we present evidence that Bax activation requires BH3-only proteins such as tBid and additional factors that are not yet characterized. Moreover, we present data about the role of proteins of the mitochondrial fission machinery in MOMP.

Bax activation requires a BH3-only protein and a cytosolic/mitochondrial factor

tBid triggers Bax insertion and oligomerization in the outer mitochondrial membrane

During apoptosis, Bax undergoes conformational changes, translocates to the MOM and oligomerizes. This series of events can be studied with isolated mitochondria using recombinant proteins (100 nM recombinant full length Bax and 10 nM tBid). Following incubation with mitochondria, Bax does not insert in the mitochondrial membrane (insertion of Bax in the membrane can be assessed by resistance to alkali treatments) and there is no cytochrome c release (release of cytochrome c can be assessed by western blotting of cytochrome c in the mitochondrial pellet and in the mitochondrial supernatant using an antibody to cytochrome c). However, when incubated in the presence of tBid, Bax inserts in the MOM, oligomerizes and this is accompanied by the release of cytochrome c (oligomerization can be studied by cross linking experiments and gel filtration analysis of proteins extracted from mitochondria with CHAPS).

tBid does not trigger Bax oligomerization in liposomes

A stepwise more reductionist approach can be used to study the mechanism of Bax activation relying on the use of synthetic liposomes made with the lipid composition of the MOM. When such liposomes were used, it was found that Bax alone or in combination with tBid was able to insert in the liposomes but failed to oligomerize, suggesting that in addition to tBid, another factor is required to allow Bax oligomerization (Roucou et al 2002). This factor, BAF (Bax activating factor), was found to be present on the mitochondrial surface and in the cytosol. In the presence of tBid, Bax was able to oligomerize on purified MOM. On the other hand, Bax was able to oligomerize in liposomes incubated with cytosol and tBid. The identity of BAF is still unknown.

Involvement of components of the permeability transition pore (PTP)

Many studies have suggested an involvement of components of the PTP such as VDAC or ANT in Bax activation. We have previously reported data against an involvement of ANT and cyclophilin D in Bax activation (Eskes et al 1998). Experiments performed in mice deficient in ANT or cyclophilin D have allowed the requirement of these proteins in Bax activation to be excluded (reviewed in Lucken-Ardjomande & Martinou 2005) and thereby have confirmed our previous data (Eskes et al 1998). Moreover, we have reported that Bax can permeabilize the outer membrane of VDAC-deficient yeast (Roucou et al 2002). It will be interesting to test the requirement of VDAC isoforms in Bax activation in mammalian cells.

In summary, we do not think that the factor that is required for Bax activation is part of the PTP. We think that it is a soluble cytosolic protein which under specific circumstances can translocate to the mitochondria.

The role of mitochondrial fission in the release of cytochrome c

Overexpression of Bax triggers mitochondrial fission

We have previously reported that the size of mitochondria decreases significantly in sympathetic neurons undergoing apoptosis after nerve growth factor (NGF) deprivation (Martinou et al 1999). In addition, we demonstrated that addition of recombinant Bax, directly to isolated mitochondria, triggered a size reduction of the organelles (Martinou et al 1999). Here we show that in HeLa cells transfected with a plasmid for Bax expression, all Bax-expressing cells display punctiform mitochondria (visualised with an antibody directed to mitochondrial Hsp70), whereas cells that do not express the plasmid show long, filamentous mitochondria (Fig. 1). Altogether, these data indicate that Bax triggers mitochondrial fission.

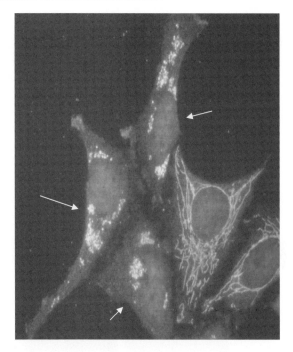

FIG. 1. Overexpression of Bax triggers mitochondrial fission. HeLa cells cultured in the presence of z-VAD, a peptide that inhibits caspases, were transfected with a plasmid encoding Bax. Twenty four hour later, the cells were fixed and stained with an antibody directed against mitochondrial Hsp70. The arrows indicate the cells that express Bax.

Inhibiting the mitochondrial fission machinery prevents the release of cytochrome c, but not that of Smac/DIABLO, and does not prevent nuclear condensation

We have examined the involvement of mitochondrial fission in the release of cytochrome c and Smac/DIABLO during apoptosis induced by UV irradiation. The release of cytochrome c and Smac/DIABLO was assessed in control and Drp1-depleted cells after UV irradiation. UV irradiation of control transfectants led to partial detachment of the cells, release of cytochrome c and Smac/DIABLO and condensation of nuclear DNA (Fig. 2). In contrast, cytochrome c staining was distinctively punctate in UV irradiated Drp1-depleted cells, while Smac/DIABLO appeared to be released, as in control cells. Western blot analysis confirmed that the release of cytochrome c was incomplete in Drp-1-deficient cells (data not shown). Importantly, Drp1-deficient cells displayed nuclear condensation (Fig. 2) as did irradiated control cells. Together, these results suggest that Drp1-depleted cells undergo apoptosis, as assessed by DNA condensation and caspase activation (not shown), without a complete release of cytochrome c from the mitochondria.

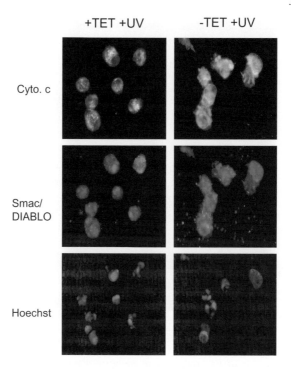

FIG. 2. Depleting cells of Drp1 partially inhibits the release of apoptogenic proteins from the mitochondria. HeLa Drp1 shRNA inducible cell line was cultured in the presence or absence of tetracycline. The cells were UV irradiated and, 14 h later, localization of cytochrome c and Smac/DIABLO was assessed by immunostaining. The cells were also co-stained with Hoechst 33342 to observe changes in the nuclear morphology.

We think that, during apoptosis, permeabilization of the MOM by Bax/Bak would trigger the release of soluble intermembrane space proteins, (such as Smac/DIABLO, HtrA2/Omi and some cytochrome c) followed by a Drp1/OPA1-dependent remodelling of the mitochondrial cristae (Scorrano & Korsmeyer 2003). This would lead to the liberation of the pool of cytochrome c located in the cisternae and/or bound to the inner mitochondrial membrane (Bernardi & Azzone 1981, Iverson & Orrenius 2004, Parone et al 2006). In the HeLa cells used in our studies, the amount of soluble cytochrome c that is released following MOMP appears to be sufficient to trigger caspase activation and apoptosis. It is possible, however, that in other cell types, the complete release of mitochondrial apoptogenic proteins is required to activate caspases and to trigger cell death.

Conclusion

In conclusion, the mechanisms of Bax activation and MOMP are still not well understood. Many theories have been postulated to explain these events. We think that the role of the mitochondrial fission machinery in the activation of Bax requires further investigations. Concerning MOMP, more and more data suggest that the PTP is not involved. On the other hand, the existence of a large pore formed by Bax alone appears difficult to reconcile with the release of a large number of proteins, the size of which can be >100 kDa. For these reasons, we favour the possibility that Bax oligomerization triggers a change in the structure of the lipid bilayer such as the formation of lipid hexagonal phases that would be unstable and highly permeable (Lucken-Ardjomande & Martinou 2005).

Acknowledgements

We are grateful to the members of the Martinou lab for stimulating discussions and critical reading of the manuscript. This work was funded by the Swiss National Science Foundation (subside: 3100A0-109419/1), OncoSuisse trust, the Medic Foundation and the Geneva Department of Education.

References

Arnoult D, Grodet A, Lee YJ, Estaquier J, Blackstone C 2005 Release of OPA1 during apoptosis participates in the rapid and complete release of cytochrome c and subsequent mitochondrial fragmentation. J Biol Chem 280:35742–35750
Bernardi P, Azzone GF 1981 Cytochrome c as an electron shuttle between the outer and inner mitochondrial membranes. J Biol Chem 256:7187–7192
Bossy-Wetzel E, Barsoum MJ, Godzik A, Schwarzenbacher R, Lipton SA 2003 Mitochondrial fission in apoptosis, neurodegeneration and aging. Curr Opin Cell Biol 15:706–716
Cory S, Adams JM 2002 The Bcl2 family: regulators of the cellular life-or-death switch. Nat Rev Cancer 2:647–656
Danial NN, Korsmeyer SJ 2004 Cell death: critical control points. Cell 116:205–219
Desagher S, Osen-Sand A, Nichols A et al 1999 Bid-induced conformational change of Bax is responsible for mitochondrial cytochrome c release during apoptosis. J Cell Biol 144:891–901
Ekert PG, Vaux DL 2005 The mitochondrial death squad: hardened killers or innocent bystanders? Curr Opin Cell Biol 17:626–630
Eskes R, Antonsson B, Osen-Sand A et al 1998 Bax-induced cytochrome c release from mitochondria is independent of the permeability transition pore but highly dependent on Mg2+ ions. J Cell Biol 143:217–224
Eskes R, Desagher S, Antonsson B, Martinou JC 2000 Bid induces the oligomerization and insertion of Bax into the outer mitochondrial membrane. Mol Cell Biol 20:929–935
Frank S, Gaume B, Bergmann-Leitner ES et al 2001 The role of dynamin-related protein 1, a mediator of mitochondrial fission, in apoptosis. Dev Cell 1:515–525
Green DR, Reed JC 1998 Mitochondria and apoptosis. Science 281:1309–1312
Huang DC, Strasser A 2000 BH3-only proteins—essential initiators of apoptotic cell death. Cell 103:839–842

Iverson SL, Orrenius S 2004 The cardiolipin-cytochrome c interaction and the mitochondrial regulation of apoptosis. Arch Biochem Biophys 423:37–46

Jagasia R, Grote P, Westermann B, Conradt B 2005 DRP-1-mediated mitochondrial fragmentation during EGL-1-induced cell death in C. elegans. Nature 433:754–760

Kuwana T, Mackey MR, Perkins G et al 2002 Bid, Bax, and lipids cooperate to form supramolecular openings in the outer mitochondrial membrane. Cell 111:331–342

Letai A, Bassik MC, Walensky LD, Sorcinelli MD, Weiler S, Korsmeyer SJ 2002 Distinct BH3 domains either sensitize or activate mitochondrial apoptosis, serving as prototype cancer therapeutics. Cancer Cell 2:183–192

Lucken-Ardjomande S, Martinou JC 2005 Newcomers in the process of mitochondrial permeabilization. J Cell Sci 118:473–483

Martinou I, Desagher S, Eskes R et al 1999 The release of cytochrome c from mitochondria during apoptosis of NGF-deprived sympathetic neurons is a reversible event. J Cell Biol 144:883–889

Martinou JC, Green DR 2001 Breaking the mitochondrial barrier. Nat Rev Mol Cell Biol 2:63–67

Martinou JC, Youle RJ 2006 Which came first, the cytochrome c release or the mitochondrial fission? Cell Death Differ 13:1291–1295

Parone P, James DI, Da Cruz S et al 2006. Inhibiting the mitochondrial fission machinery does not prevent Bax/Bak-dependent apoptosis. Mol Cell Biol 26:7397–7408

Pitts, KR, Yoon Y, Krueger EW, McNiven MA 1999 The dynamin-like protein DLP1 is essential for normal distribution and morphology of the endoplasmic reticulum and mitochondria in mammalian cells. Mol Biol Cell 10:4403–4417

Roucou X, Montessuit S, Antonsson B, Martinou JC 2002 Bax oligomerization in mitochondrial membranes requires tBid (caspase-8-cleaved Bid) and a mitochondrial protein. Biochem J 368:915–921

Rube DA, van der Bliek AM 2004 Mitochondrial morphology is dynamic and varied. Mol Cell Biochem 256–257:331–339

Scorrano L, Korsmeyer SJ 2003 Mechanisms of cytochrome c release by proapoptotic BCL-2 family members. Biochem Biophys Res Commun 304:437–444

Smirnova E, Shurland DL, Ryazantsev SN, van der Bliek AM 1998 A human dynamin-related protein controls the distribution of mitochondria. J Cell Biol 143:351–358

Terradillos O, Montessuit S, Huang DC, Martinou JC 2002 Direct addition of BimL to mitochondria does not lead to cytochrome c release. FEBS Lett 522:29–34

Vander Heiden MG, Thompson CB 1999 Bcl-2 proteins: regulators of apoptosis or of mitochondrial homeostasis? Nat Cell Biol 1:E209–E216

Youle RJ, Karbowski M 2005 Mitochondrial fission in apoptosis. Nat Rev Mol Cell Biol 6:657–663

DISCUSSION

Scorrano: Another crucial lipid that is always missing from the liposomes is cholesterol. This induces phase separation in the membrane; it is a major constituent of the OMM. Whenever we want to model this in an acellular system, we should always include cholesterol. Have you looked at whether with cholesterol you get oligomerization of Bax independently of the cytosolic factor?

Martinou: We have added cholesterol to these membranes. It has a tendency to prevent the oligomerization of Bax.

Orrenius: If cardiolipin is necessary to create these domains for Bax insertion and oligomerization, and Baf is instrumental, this would mean that Baf would first have to move cardiolipin from the inner membrane to the outer membrane. Is this the action of Baf, acting on the lipid rather than on Bax?

Martinou: We don't know for sure whether there is cardiolipin on the outer membrane. Some papers claim this is the case (Feo et al 1973); if so, we don't need to move it from inside to out.

Orrenius: On the other hand, a number of reports claim (see Orrenius et al 2007 for review) that under normal conditions the bulk of cardiolipin is in the inner membrane. If you want to target cardiolipin from the outside of the mitochondria, you'll probably need to go to contact sites.

Martinou: I agree with you. I am always surprised when I see these numbers, such as claims that 25% of cardiolipin is on the outer leaflet of the outer membrane: I don't know how people come up with these figures because it is so difficult to purify these membranes. Let's assume that cardiolipin is in the outer membrane. In this case, the proteins would have direct access to it. If it isn't, then we have to think of a mechanism for putting it there. Baf or Bid could be responsible.

Jacobs: Can the oligomerization of Bax be reversed? Does oligomerized Bax oligomerize more non-oligomerized Bax on its own? And what is the evidence that this protease-resistant form really is an oligomer and not just a protease resistant form?

Martinou: If we analyse the Bax on gel filtration following the addition of Bid and Baf, it is not eluted in the low molecular fraction but is in the high molecular fraction. We think it has a different structure. We can't explain this by micelles, for example.

Giulivi: What happens if you run this high molecular weight aggregate in SDS PAGE under reducing, denaturing conditions?

Jacobs: It falls apart.

Martinou: Yes, the oligomer is not resistant to denaturing conditions.

Lemasters: Sten Orrenius has been interested in oxidation of cardiolipin (Iverson & Orrenius 2004). Does oxidation of cardiolipin make a difference to the extent of oligomerization?

Martinou: We haven't examined this, but it is something that we should do. Perhaps it is naturally oxidised in our conditions.

Youle: We found that if you knockout endophilin B1, long tubules stem from the outer membrane. It sounds like your Baf. If we take pure endophilin and add Bax, it induces endophilin B1 to form oligomers (our unpublished observations).

Shirihai: You suggested that Bax reduced the size of mitochondria. What methodology did you use?

Martinou: This was by electron microscopy. We analysed serial cross sections of mitochondria.

Hajnóczky: How did you separate shrinkage from fission? Did you count the number of mitochondria per cell?

Martinou: We didn't look at the volume of the mitochondria.

Shirihai: Do you treat fission and fragmentation as identical processes?

Martinou: Yes, I used two words for the same process.

Shirihai: I'd distinguish them: fission is defined as the process that breaks a mitochondrion with a continuous matrix lumen into two mitochondria that have separated matrixes. Not every fission is accompanied by movement of the two daughter mitochondria away from each other. Fragmentation defines the appearance of mitochondria in light microscopy as multiple mitochondria that are not juxtaposed. Juxtaposed mitochondria are not necessarily fused and therefore can become fragmented without going through fission.

Reynolds: In the pictures you showed of Bax overexpression, there were a couple of cells with long extended tubules all through the cell, and then there were little punctate mitochondria. These punctate mitochondria weren't all over the cell; they were in a little spot at the side of the nucleus. Maybe the fragmentation is just part of the process needed to get the mitochondria to that spot in the cell where they are going to do damage.

Martinou: That is possible. Is fission/fragmentation necessary for apoptosis to occur? Based on data mainly obtained in Youle's lab, we assume that fission plays an important role, since depletion of key proteins involved in the fission/fragmentation machinery, leads to cell protection against various stimuli. We have done similar experiments in which we have depleted the cells of Drp1 or Fis1. Removal of these proteins by RNAi treatment results in cells with elongated mitochondria. Following UV radiation, cytochrome c is maintained in mitochondria but Smac is released. Have we protected the cells? No, they aren't significantly protected.

Youle: That is quite interesting, considering the lipidic pore model that would not be predicted to distinguish between soluble inter-membrane space proteins to be released. One simple hypothesis is that Drp1 is inhibiting cardiolipin oxidation, and this is why the cytochrome c is being retained and not Smac/DIABLO.

Schon: There is a reagent you might not be aware of. It is a monoclonal antibody against cardiolipin.

Martinou: I have talked to the people working on diseases in which people develop antibodies to cardiolipin. In fact, the antibodies don't recognize cardiolipin: they recognize cardiolipin bound to a glycoprotein.

Schon: The paper I am referring to (Haynes et al 2005) involved making monoclonals to HIV proteins, and they discovered that one of them was a monoclonal against cardiolipin. Another issue is that I don't think that humanin exists.

Scorrano: Gordon Shore's group showed that in response to the activation of Drp1, mitochondria do remodel their cristae (Germain et al 2005). This could go into the proposed mechanism that you just showed.

Bernardi: This is a good point. It means that in the absence of Drp1 there is no remodelling of the cristae.

Martinou: In that setting, yes.

Bernardi: I think the remodelling is redundant. We need to permeabilize the outer membrane.

Youle: Do you think your Baf is a Drp1 binding protein?

Martinou: It is not impossible.

Scorrano: Timing is always difficult during these cascades, but my idea has always been that there should be a threshold of cytochrome c which should be reached in the cytosol. This has been clearly demonstrated by microinjection of cytochrome c. If you don't inject a threshold level the cell won't die. This could potentially be important in the release of the whole pool of cytochrome c. In tumour cells where there is overexpression of the inhibitory proteins of caspases and this needs to be overcome, releasing all cytochrome c could help. You don't know the threshold of cytochrome c release required in your particular cell line that you use in that experiment to drive apoptosis. If you could not reach that threshold, then the cells wouldn't undergo apoptosis. I never said that cristae remodelling is essential for driving the progression of apoptosis. We only see a kinetic issue here, i.e. delaying of apoptosis when we block cristae remodelling. We are dealing with a complex phenomenon in living tissues. We don't know whether slowing down a process such as apoptosis is sufficient to cause something that is bad for the whole organism.

Martinou: Indeed, caspases could be activated by low amounts of cytochrome c in the cells that we have used.

Halestrap: We talked yesterday about uncoupler fragmenting mitochondria. If uncoupler is added with oligomycin in many cells ATP can be maintained. We know that these cells undergo apoptosis: is that because the fragmentation is releasing cytochrome c? If you add uncoupler under energized conditions, is cytochrome c released? This would be fragmentation without any involvement of Bax, and I am interested to know what happens.

Martinou: I should have mentioned that we are studying Bax/Bak-dependent apoptosis. Cytochrome c can be released in the absence of activation of Bax and Bak, in which case the PTP could be involved. What I have said is true for Bax and Bak-dependent apoptosis.

Halestrap: If you fragment and don't open the PTP pore do you get cytochrome c release and apoptosis?

Martinou: Fragmentation alone doesn't lead to apoptosis. Some cells can live perfectly well with fragmented mitochondria. What I say is that we have never been able to find elongated mitochondria depleted of cytochrome c in Bax/Bak-dependent apoptosis. Each time we see a mitochondrion depleted of cytochrome c, it is fragmented.

Jacobs: If different cell types vary in their sensitivity to cytochrome c release because they are either poised or not poised to go into apoptosis, then fragmentation alone may release a small amount of cytochrome c in many conditions that have nothing to do with Bax and Bak. In many cell types this does not then trigger apoptosis because there is just not enough of it. However, under some specific conditions it might suffice.

Nicholls: It is easier if we stick to the term fission, rather than use this interchangeably with fragmentation. The question is, is there a risk that normal physiological fission releases a little bit of cytochrome c?

Halestrap: Isn't there a paper that says that Bax migrates to mitochondria when you add uncoupler (Smaili et al 2001)?

Bernardi: That was pore opening driving Bax translocation (Giorgio et al 2002).

Scorrano: If you treat Bax/Bak double knockout cells with staurosporine or etoposide, the mitochondria do fragment. Yet at this point they still haven't released their cytochrome c. I would not say that fission occurs via the release of cytochrome c in that model.

Shirihai: I want to clarify an experimental point. If one looks at a cell with mitochondria that appear more connected (juxtaposed), and after a treatment the mitochondria are becoming more fragmented, one cannot tell that fission has occurred. The transition might have occurred on the cytoskeletal level, moving unfused, juxtaposed mitochondria away from each other, an event not related to fission. This is the observation of fragmentation; the observation of fission can be without fragmentation.

Nicholls: So we are talking about physical separation rather than continuity.

O'Rourke: You discounted the Bax pore model. Can you comment on the Casey Kinnally data (Dejean et al 2006) showing that there are channels in the outer membrane in cells that are induced to apoptose, and there are more of them in the presence of Bax than in the Bax knockout?

Martinou: Are you are talking about experiments in cells?

O'Rourke: This is looking at mitochondria isolated from cells that were induced to apoptose. They have higher levels of this so-called MAC channel.

Martinou: I am aware of these results and I acknowledge that she is able to measure channel activity. Can she exclude that it is a lipidic pore?

O'Rourke: Why would you exclude the formation of a proteinaceous pore?

Martinou: Rapidity of release, and the release of very large proteins which would necessitate a large pore.

Scorrano: One of the problems in interpreting these data from Kinnally is that the conductance was not inhibited by the addition of cytochrome c. She claimed that this was her biggest interpretational problem. Then if we go back to the lipidic story, personally I am concerned by the model put out by Newmeyer's group

(Kuwana et al 2002). They see that across this putative lipidic pore, molecules of 200 kDa and 2000 kDa diffuse at the same rate. I have a biophysical problem understanding how this could happen. I would expect a 2000 kDa protein to diffuse slower. It could be a break in the outer mitochondrial membrane rather than a pore.

Martinou: We have to be careful with permeabilization of liposomes.

Bernardi: Can we go back to the issue of cardiolipin. What kind of molar ratio of cardiolipin do you need in your liposomes? If it is 25% you have a problem.

Martinou: We have tried different concentrations. We need a huge amount: 20%.

Bernardi: You have a problem, then: there's no way you can explain this being in the outer membrane normally.

Martinou: We still don't understand why we need so much cardiolipin in our acellular assay.

Youle: On the issue of Drp1 dependence on cell death, there is a new paper looking at three different alleles of Drp knockout or mutation in the fly (Goyal et al 2007) showing a robust inhibition of programmed cell death. This could help sort this issue out. There exist papers saying there is no cytochrome c release in the fly, so what is Drp1 doing in the fly?

Orrenius: With regards to cardiolipin, most of the work has been done with yeast. We looked at cardiolipin synthase mutants in yeast, and we could permeabilize the mitochondria and release cytochrome c from such mutants with oligomeric Bax (Iverson et al 2004).

Schon: Even in the CS-minus?

Orrenius: Yes. On the other hand, because of the down-regulation of the cardiolipin synthase, the cardiolipin precursor, phosphatidylglycerol, accumulates in large amounts in these membranes, and they also become leakier to cytochrome c compared to control mitochondria.

Martinou: We have tested whether phosphatidylglycerol can replace cardiolipin in the acellular assay we developed. It doesn't work.

Halestrap: As I understand it, after cytochrome c is released you can rescue cells and the mitochondria will start functioning normally. Presumably, they have somehow sealed their outer membranes. Whatever this big pore is, it needs to be reversible.

Martinou: It is astonishing that mitochondria can reform normally with a regular cytochrome c content.

Nicholls: Are we sure they are the same mitochondria and not fresh ones?

Youle: This could explain the results. If not all the mitochondria release their cytochrome c, then the ones that don't could repopulate the cell.

Nicholls: This could be a good model for biogenesis.

Martinou: Yes, and protein synthesis is needed to restore the function.

References

Dejean LM, Martinez-Caballero S, Kinnally KW 2006 Is MAC the knife that cuts cytochrome c from mitochondria during apoptosis? Cell Death Differ 13:1387–1395

Feo F, Canuto RA, Bertone G, Garcea R, Pani P 1973 Cholesterol and phospholipid composition of mitochondria and microsomes isolated from morris hepatoma 5123 and rat liver. FEBS Lett 33:229–32

Germain M, Mathai JP, McBride HM, Shore GC 2005 Endoplasmic reticulum BIK initiates DRP1-regulated remodelling of mitochondrial cristae during apoptosis. EMBO J 24:1546–56

Giorgio M, Migliaccio E, Orsini F et al 2005 Electron transfer between cytochrome c and p66Shc generates reactive oxygen species that trigger mitochondrial apoptosis. Cell 122:221–233

Goyal G, Fell B, Sarin A, Youle RJ, Sriram V 2007 Role of mitochondrial remodeling in programmed cell death in Drosophila melanogaster. Dev Cell 12:807–816

Haynes BF, Fleming J, St Clair EW et al 2005 Cardiolipin polyspecific autoreactivity in two broadly neutralizing HIV-1 antibodies. Science 308:1906–1908

Iverson SL, Orrenius S 2004 The cardiolipin-cytochrome c interaction and the mitochondrial regulation of apoptosis. Arch Biochem Biophys 423:37–46

Iverson SL, Enoksson M, Gogvadze V, Ott M, Orrenius S 2004 Cardiolipin is not required for Bax-mediated cytochrome c release from yeast mitochondria J Biol Chem 279:1100–1107

Kuwana T, Mackey MR, Perkins G et al 2002 Bid, Bax, and lipids cooperate to form supramolecular openings in the outer mitochondrial membrane. Cell 111:331–342

Orrenius S, Gogvadze V, Zhivotovsky B 2007 Mitochondrial oxidative stress: implications for cell death. Annu Rev Pharmacol Toxicol 47:143–183

Smaili SS, Hsu YT, Sanders KM, Russell JT, Youle RJ 2001 Bax translocation to mitochondria subsequent to a rapid loss of mitochondrial membrane potential. Cell Death Differ 8:909–920

Mitochondria and neurodegeneration

M. Flint Beal

Weill Medical College of Cornell University, Department of Neurology and Neuroscience, 525 East 68th Street, New York, NY 10021, USA

> *Abstract.* There is increasing evidence linking mitochondrial dysfunction to neurodegenerative diseases. Mitochondria are critical regulators of cell death, a key feature of neurodegeneration. Mutations in mitochondrial DNA and oxidative stress both contribute to ageing, which is the greatest risk factor for neurodegenerative diseases. This is the case in Alzheimer's disease, in which there is evidence that both β-amyloid and the amyloid precursor protein may directly interact with mitochondria, leading to increased free radical production. In the case of Huntington's disease (HD), recent evidence suggests that the coactivator PGC1α, a key regulator of mitochondrial biogenesis in respiration, is down-regulated in patients with HD and in several animal models of this neurodegenerative disorder. In Parkinson's disease, the autosomal recessive genes parkin, DJ1 and PINK1 are all linked to either oxidative stress or mitochondrial dysfunction. In amyotrophic lateral sclerosis, there is strong evidence that mutant superoxide dismutase directly interacts with the outer mitochondrial membrane as well as the intermembrane space and matrix. Therefore, an impressive number of disease specific proteins interact with mitochondria. Therapies that target basic mitochondrial processes such as energy metabolism in free radical generation, or specific interactions of disease-related protein with mitochondria, hold great promise.
>
> *2007 Mitochondrial biology: new perspectives. Wiley, Chichester (Novartis Foundation Symposium 287) p 183–196*

Neurodegenerative diseases are characterized by gradually progressive selective loss of anatomically or physiologically related neuronal systems. Amongst these diseases are Alzheimer's disease (AD), Parkinson's disease (PD) amyotrophic lateral sclerosis (ALS) and Huntington's disease (HD). There is increasing evidence that mitochondrial dysfunction plays an important role in the pathogenesis of these diseases (Lin & Beal 2006). Mitochondria are key regulators of cell survival and death. They have a central role in ageing and have recently been found to interact with many of the specific proteins that are implicated in genetic forms of neurodegenerative diseases. The most important risk factor for neurodegenerative diseases such as AD, PD and ALS is ageing. Mitochondria are thought to contribute to ageing through the accumulation of mitochondria DNA (mtDNA) mutations and increased production of reactive oxygen species. It is well established that mtDNA mutations accumulate with normal ageing. This is especially the case with large-scale deletions and point mutations. We found that the accumulation of point mutations in the cytochrome oxidase gene correlates with a reduction in the activity of this enzyme (Lin et al 2002). Others have found high levels of mtDNA deletions in the individual neurons in the substantia nigra (Bender et al 2006). These show an age-dependent accumulation particularly in cytochrome oxidase-deficient neurons. This may well contribute to the age-dependence of PD. In mice which have an impairment of the proofreading part of the mtDNA polymerase, there is a marked increase in mtDNA point mutations and deletions (Trifunovic 2006). This is because the mtDNA mutations accumulate due to uncorrected errors during replication. These mice show up to a ninefold increase in point mutations in cytochrome b. This results in decreased respiratory enzyme activity in ATP production. The mice show a marked increase in age-related phenotypes such as weight loss, alopecia, osteoporosis, kyphosis, cardiomyopathy, anaemia, skin atrophy and sarcopenia. The median lifespan of such mice is 48 weeks, which is much shorter than the typical murine lifespan of two years. The mice show increased markers of apoptosis; yet, there does not appear to be oxidative damage to lipids, proteins or DNA.

AD is characterized clinically by a progressive decline in cognitive abilities, and pathologically by the presence of senile plaques composed primarily of amyloid β peptide. There are also neurofibrillary tangles made up of hyperphosphorylated tau. In autosomal dominant inherited disease, three proteins have been associated, including the amyloid precursor protein (APP), which is cleaved sequentially by β and γ secretases to produce amyloid β (Aβ) and presenilins 1 and 2, which are components of the γ secretase complex.

In AD, there appears to be increased oxidative damage, which precedes the onset of significant plaque pathology. It also precedes Aβ deposition in transgenic APP mice (Pratico et al 2001). Oxidative damage appears to increase production of intracellular Aβ. This has been shown both in cell culture using fetal guinea pig neurons, as well as in cultured human astrocytes treated with the mitochondrial uncoupler CCCP. We found that transgenic APP mutant mice crossed with hemizygous mice with a deficiency of the antioxidant enzyme mnSOD, showed markedly increased brain Aβ levels in plaque deposition (Lin et al 2004). Similar findings have been observed in mice that are deficient in the vitamin E binding protein. This leads to increased oxidative stress and increased Aβ plaque generation (Nishida et al 2006). In another transgenic APP mutant mouse, energy metabolism inhibitors including insulin, 2-deoxyglucose and 3-nitropropionic acid, all increased β secretase levels as well as Aβ levels. Oxidative stress increases the expression of β secretase through activation of c-Jun N-terminal kinase and p38 mitogen activated protein kinase. It also results in increased aberrant tau phosphorylation by activation of glycogen synthase kinase 3. Oxidation also inactivates the prolyl isomerase PIN1 (Pastorino et al 2006). PIN1 catalyses protein conformational changes that can increase both APP and tau processing.

There are several recent reports that many of the proteins implicated in AD pathogenesis have a direct physical involvement with mitochondria or mitochondrial proteins. APP has been linked to dysfunction of mitochondria (Devi et al 2006). It has been shown that APP carries a dual leader sequence permitting targeting to the endoplasmic reticulum or to mitochondria. When overexpressed in cultured cells, APP is found in the mitochondrial enriched cell fractions and APP immunoreactivity was seen in mitochondria by immunoelectron microscopy. The mitochondrial association is not seen if the leader sequence is mutated. Using chemical cross-linkers, APP was shown to be in contact with the mitochondrial protein importation machinery. It was shown that a large acidic domain spanning APP residues 220–290 caused APP to become stuck during importation with the N-terminus inside, and the C-terminus outside. Accumulation of this transmembrane arrested APP was associated with reduced cytochrome oxidase activity, decreased ATP synthesis and loss of mitochondrial membrane potential. When the acidic APP 220–290 domain was deleted, transmembrane arrest did not occur, and mitochondrial function was not impaired. This work has been extended to post-mortem brain samples from human AD subjects and control subjects. Unglycosylated full length and C-terminally truncated APP was associated with mitochondria in samples from the brains of individuals with AD, but not in mitochondria in samples from subjects without the disease. In the AD brain samples, levels of mitochondrial APP were higher in affected brain regions and in subjects with more advanced disease. Without resorting to chemical cross-linking, the authors showed using both blue native gels and immunoelectron microscopy that APP was stably associated with two components in the mitochondrial protein translocation machinery, TOM40 and TIM23 (translocases of the outer and inner membranes). This provided further evidence suggesting that APP clogs this machinery. Consistent with this, higher mitochondrial APP levels *in vitro* were associated with decreased importation of respiratory gene subunits including decreased cytochrome oxidase activity and resulted in H_2O_2 generation.

Aβ has also been reported to be found within the mitochondrial matrix where it binds to the enzyme termed the Aβ binding alcohol dehydrogenase (ABAD) (Lustbader et al 2004). Blocking the interaction of Aβ with ABAD with a decoy peptide suppressed Aβ-induced apoptosis and free radical generation in neurons. Conversely, overexpression of ABAD in transgenic APP mice exaggerated neuronal oxidative stress and impaired memory. Other groups have also reported that Aβ interacts with mitochondria inhibiting cytochrome oxidase activity and increasing free radical generation. Aβ also inhibits α-ketoglutarate dehydrogenase activity in mitochondria and a deficiency of α-ketoglutarate dehydrogenase has been consistently found in postmortem brain tissue of AD patients (Gibson et al 1988).

A role of mitochondria in the pathogenesis of PD has been strongly suggested by toxin models using MPTP (1-methyl-4-phenyl-1,2,3,6-tetrahydropyridine), as well as with rotenone, both of which result in inhibition of complex I of the electron transport chain. Both of these toxins can produce a parkinsonian phenotype in experimental animals. Complex I inhibition and oxidative stress were shown to be relevant in naturally occurring PD in which complex I deficiency and glutathione depletion are observed in the substantia nigra of patients.

Several recent genetic observations have strongly implicated mitochondria in PD pathogenesis. Mutations or polymorphisms in mtDNA, as well as nuclear genes have been identified as causing PD. Of the nuclear genes, α-synuclein, parkin, DJ1, PINK1, LRRK2 and HTRA2 directly or indirectly involve mitochondria (Klein & Schlossmacher 2006).

A small number of cases of PD have been associated with mtDNA mutations. Several groups have found that certain continent specific clusters of polymorphisms, termed mtDNA haplotypes, may decrease the risk of developing PD. Among Europeans, the haplotype cluster UJKT is associated with a reduced risk for PD, as compared with haplotype H (Pyle et al 2005). These same haplotypes, which are underrepresented in PD patients, are overrepresented in healthy centenarians. It has been hypothesized that the haplotypes may lead to partial uncoupling of mitochondria and, thereby, increase longevity and decrease the risk for neurodegeneration by reducing free radical generation. Mutations in α-synuclein are associated with autosomal dominant familial PD. α-synuclein is a major component of Lewy bodies, and the primary effect of the mutations appears to be formation of oligomeric or fibrilloaggregates. In transgenic mice, overexpression of α-synuclein impairs mitochondrial function, increases oxidative stress and enhances nigral pathology induced by MPTP. A recent study of mice overexpressing A53T mutant α-synuclein showed degenerating mitochondria that immunostained for α-synuclein, raising the possibility that mutant α-synuclein may damage mitochondria directly (Martin et al 2006). Whereas overexpression of α-synuclein increases sensitivity to MPTP, the α-synuclein null mice are resistant to MPTP, as well as to other mitochondrial toxins.

Mutations in three genes are associated with autosomal recessive early-onset PD (Klein & Schlossmacher 2006). These are mutations in parkin, DJ1 and PINK1. It has been suggested that these proteins may interact. Parkin null *Drosophila* and mouse strains exhibit mitochondrial impairment and increased oxidative stress. Parkin can associate with the outer mitochondrial membrane and prevent mitochondria swelling, cytochrome c release and caspase activation. Parkin has also been localized to mitochondria in proliferating cells, where it has been shown to associate with the mitochondrial transcription factor A and to enhance mitochondria biogenesis. In *Drosophila*, parkin depletion results in marked mitochondrial swelling and damage, and overexpression of glutathione-*S*-transferase, which has a role in detoxifying products of oxidative damage, suppresses neurodegeneration (Whitworth et al 2005).

Mutations in DJ1 are also associated with autosomal recessive juvenile PD. The overall function of DJ1 seems to be to protect against cell death, especially that induced by oxidative stress (Bonifati et al 2003). DJ1-deficient mice are hypersensitive to MPTP and oxidative stress (Kim et al 2005). Moreover, in flies, DJ1 undergoes progressive oxidative inactivation with ageing, which in turn increases sensitivity to oxidative stress, and could contribute to the age-dependence of sporadic PD (Meulener et al 2006).

Mutations in PINK1 present a third form of autosomal recessive juvenile PD. PINK1 is a kinase localized to mitochondria, and like DJ1, seems to protect against cell death (Valente et al 2004). Overexpression of wild-type PINK1 protects against apoptosis induced by staurosporin. In *Drosophila* PINK1 deficiency causes mitochondrial pathology, increases sensitivity to paraquat and rotenone, and degeneration of flight muscles in dopaminergic neurons. This pathology is identical to that seen in parkin mutant flies and can be rescued by overexpression of parkin but not DJ1 (Yang et al 2006). Thus, PINK1 probably functions in the same pathway as parkin, with parkin downstream of PINK1. In *Drosophila* that have an inactivation of PINK1, antioxidants protect against progressive loss of dopaminergic neurons and against ommatidyl degeneration in the compound eye (Wang et al 2006). Expression of human SOD1 or treatment with the antioxidant vitamin E inhibits degeneration, suggesting that oxidative stress plays an important role in the neuronal degeneration.

There is strong evidence that there is mitochondrial impairment in ALS. ALS is characterized by progressive weakness and atrophy, as well as spasticity reflecting degeneration of the upper and lower motor neurons within the cerebral cortex, brainstem and spinal cord. Postmortem tissue, as well as biopsy samples of a variety of tissues in both sporadic ALS patients, as well as familial ALS patients, show mitochondrial abnormalities. Much of the research however, has focused on patients and mouse models involving mutations in copper/zinc superoxide dismutase (SOD). Mice overexpressing the G93A or the G37R mutations show mitochondrial vacuolation. There is also impaired calcium loading capacity in mitochondria from the brain and spinal cord, but not the liver of mice overexpressing both the G93A and the G85R SOD1 mutations. Recently, it has been shown that as many as 12 different SOD1 mutations are associated with mitochondria *in vitro* (Ferri et al 2006).

Mutant SOD1 accumulates and aggregates on the outer mitochondrial membrane and where it may clog the protein importation machinery, eventually resulting in mitochondrial dysfunction (Lin et al 2004). Mutant SOD1 has also been reported to bind to and aggregate with cytosolic heat shock proteins, and with mitochondrial

Bcl2 rendering them unavailable for antiapoptotic functions (Pasinelli et al 2004). Our own studies found that there was mutant SOD1, as well as aggregated SOD1, not only on the outer mitochondrial membrane, but also in the intermembrane space and the matrix of mitochondria (Vijayvergiya et al 2005).

There is substantial evidence implicating mitochondrial dysfunction in HD (Lin & Beal 2006). HD is characterized clinically by chorea, psychiatric disturbance, dementia, and pathologically by loss of long projection neurons in the cerebral cortex in striatum. HD is inherited in an autosomal dominant manner, and is due to an expansion of a CAG trinucleotide repeat in the huntingtin gene.

A number of different lines of evidence have shown that there is impairment of mitochondrial function in HD. One of the first indications that bioenergetic defects might be implicated in HD pathogenesis came from the finding that HD patients display pronounced weight loss despite increased caloric intake. Imaging with positron emission tomography showed a marked reduction in glucose metabolism in the basal ganglia. Nuclear magnetic resonance spectroscopy showed increased lactate in the cortex and basal ganglia. In addition, biochemical studies show reduced activity of several key components of oxidative phosphorylation including complexes II and III of the electron transport chain. There also appears to be marked reductions in aconitase activity. Studies on lymphoblasts from HD patients show decreased membrane resting potential and impaired calcium iron homeostasis. Also, marked morphological abnormalities have been observed, including derangement of the mitochondrial matrix and cristae. Some of these effects may be due to direct involvement of huntingtin with mitochondria, since many of the mitochondrial abnormalities including decreased calcium uptake capacity, can be recapitulated in normal lymphoblasts treated with a fusion protein composed of a peptide containing a pathogenic polyglutamine tract (Panov et al 2002). Neurons cultured from full-length HD mouse models, as well as from cultured striatal neurons expressing a knock-in of 111 CAG repeats into the mouse gene, showed that the mutant protein associated directly with the outer mitochondrial membrane. The knock-in mouse line also showed reduced rates of oxygen consumption and ATP production, as well as increased sensitivity to 3-nitropropionic acid. There is also reduced calcium uptake capacity.

Although the pathogenesis of these defects could be related to direct interaction of mutant huntingtin with the mitochondrial outer membrane, there is also substantial evidence that there are abnormalities in transcription, which play a role in HD pathogenesis. It has been proposed that aberrant transcriptional regulation of nuclear encoded mitochondrial genes may be involved in HD pathogenesis. Indeed, HD binds to several key transcription factors including SP1, TAF2 130 and CREB-binding protein and down-regulates their activity. Recently, three different studies have provided evidence that mutant huntingtin may impair the function of PGC1α, a key regulator of mitochondrial biogenesis and respiration. PGC1α was discovered a decade ago and has been implicated in energy homeostasis, adaptive thermogenesis, β oxidation of fatty acids and glucose metabolism. It was originally identified as a peroxisome proliferator-activated receptor γ (PPARγ)-interacting protein in brown adipose tissue.

Initial indications that PGC1α might play a role in HD pathogenesis came from studies on mice lacking PGC1α (Lin et al 2004). These mice developed spongiform degeneration predominantly in the striatum, and also showed abnormalities in brown adipose tissue, as well as hyperactivity. More recently, it has been shown that PCG1α is a potent suppressor of reactive oxygen species (ROS) and induces production of ROS scavenging enzymes (St. Pierre et al 2006). In response to treatment with the oxidative stressor hydrogen peroxide, there is a sixfold increase in PGC1α expression levels. Concomitantly, there is increased expression of genes encoding ROS defence enzymes including SOD1, SOD2, catalase and glutathione peroxidase. When cell lines were treated with RNAi against PGC1α, these increases in the antioxidant enzymes following exposure to hydrogen peroxide were blocked. Analysis of mice lacking PGC1α show that their basal expression of SOD1, SOD2 and catalase is considerably lower in the brains and heart, regions known to be very sensitive to oxidative stress. Mice which are deficient in PGC1α are much more sensitive to the effects of two neurotoxins, MPTP, a complex 1 inhibitor that produces parkinsonism as well as kainic acid, a glutamate receptor agonist that induces excitotoxicity in the hippocampus.

Further implicating PGC1α in HD are the findings of two groups. The first of these (Cui et al 2006) showed that there was impaired PGC1α levels in striata of post mortem HD patient brains, striata from a HD knock-in mouse model that overexpresses mutant huntingtin and in a cultured HD striatal cell line. There is a marked reduction in expression of PGC1α mRNA in these three sources of striatal neurons. It was shown that mutant huntingtin interferes with the formation of a CREB/TAF4 complex that regulates transcription of the gene encoding PGC1α. Furthermore, the cultured HD striatal cell line showed reduced expression of mitochondrial gene targets of PGC1α including cytochrome c and cytochrome oxidase 4. Of particular interest is the knockout HD mice with a 140 CAG repeats inserted into the murine huntingtin gene. The expression of PCG1α was reduced severalfold in medium spiny neurons, but was increased in interneurons which are spared in HD. The authors also showed that overexpression of PGC1α using a lentiviral vector produced some sparing of neuronal shrinkage in the striatum of both an N-terminal fragment HD mouse model, as well as the knock-in mouse model.

Another study also showed impaired PGC1α function in HD patient postmortem brain tissue (Weydt et al 2006). The authors found reduced expression of 24 out of 26 PGC1α target genes. They also found reduced PGC1α mRNA expression in the striatum of a transgenic mouse model of HD.

They found that there were phenotypic changes consistent with impaired PGC1α function in the HD transgenic mice. It is known that PGC1α is important in the expression of UCP1 in brown fat, which uncouples respiration resulting in heat production. Weydt et al (2006) report marked hypothermia at baseline and following cold exposure in two N-terminal transgenic mouse models of HD. On cold exposure, UCP1 expression was decreased in the brown adipose tissue from one of the HD mouse lines relative to wild-type animals implicating impaired PGC1α function in these mice. Failure to induce expression of UCP1, another PGC1α target gene, was further demonstrated in premier brown adipocytes from the N171-82Q mice. There was also evidence of a reduced ATP to ADP ratio and reduced numbers of mitochondria, similar to findings in PGC1α-deficient mice. The brown adipose tissue also showed abnormal vacuolization consistent with observations in PGC1α-deficient mice. These findings taken together, strongly suggest that impaired gene transcription, which affects mitochondrial biogenesis may play a key role in HD pathogenesis.

Summary

There is increasing evidence that mitochondrial dysfunction plays a key role in the pathogenesis of common neurodegenerative diseases. It has also been strongly implicated in other less common neurodegenerative diseases such as Friedreich's ataxia, neurodegeneration of brain iron accumulation and optic atrophy type 1. There are many important questions that remain to be answered. The findings of mitochondrial dysfunction in oxidative damage however, suggest that therapies that directly address these impairments may be useful in treatment. A number of transcription factors that increase expression of ROS scavenging enzymes may be useful. This includes particularly the NRF2/ARE and PGC1α signalling pathways.

References

Bender A, Krishnan KJ, Morris CM et al 2006 High levels of mitochondrial DNA deletions in substantia nigra neurons in aging and Parkinson's Disease. Nat Genet 38:515–517

Bonifati V, Rizzu P, Squitieri F et al 2003 Mutations in the DJ-1 gene associated with autosomal recessive early-onset parkinsonism. Science 299:256–259

Cui L, Jeong H, Borovecki F, Parkhurst CN, Tanese N, Krainc D 2006 Transcriptional repression of PGC-1 alpha by mutant huntingtin leads to mitochondrial dysfunction and neurodegeneration. Cell 127:59–69

Devi L, Prabhu BM, Galati DF, Avadhani NG, Anandatheerthavarada HK 2006 Accumulation of amyloid precursor protein in the mitochondrial import channels of human Alzheimer's disease brain is associated with mitochondrial dysfunction. J Neurosci 26:9057–9068

Ferri A, Cozzolino M, Crosio C 2006 Familial ALS-superoxide dismutases associate with mitochondria and shift their redox potentials. Proc Natl Acad Sci USA 103:13860–13865

Gibson GE, Sheu KF, Blass JP et al 1988 Reduced activities of thiamine-dependent enzymes in the brains and peripheral tissues of patients with Alzheimer's disease. Arch Neurol 45: 836–840

Kim RH, Smith PD, Aleyasin H et al 2005 Hypersensitivity of DJ-1 deficient mice to 1-methyl-4-phenyl-1, 2, 3, 6-tetrahydropyridine (MPTP) and oxidative stress. Proc Natl Acad Sci USA 102:5215–5220

Klein C, Schlossmacher MG 2006 The genetics of Parkinson's disease: implication for neurological care. Nat Clin Pract Neurol 2:136–146

Lin MT, Beal MF 2006 Mitochondrial dysfunction and oxidative stress in neurodegenerative diseases. Nature 443:778–795

Lin MT, Simon DK, Ahn C, Kim LM, Beal MF 2002 High aggregate burden of somatic mtDNA point mutations in aging and Alzheimer's disease brain. Hum Mol Genet 11:133–145

Lin J, Wu PH, Tarr PT et al 2004 Defects in adaptive energy metabolism with CNS-linked hyperactivity in PGS-1 alpha null mice. Cell 119:121–135

Lustbader JW, Cirilli M, Lin C et al 2004 ABAD directly links Aβ to mitochondrial toxicity in Alzheimer's Disease. Science 304:448–452

Martin LJ, Pan Y, Price AC et al 2006 Parkinson's disease α-synuclein transgenic mice develop neuronal mitochondrial degeneration and cell death. J Neurosci 26:41–50

Meulener MC, Xu K, Thompson L, Ischiropoulos H, Bonini NM 2006 Mutational analysis of DJ-1 in *Drosophila* implicates functional inactivation by oxidative damage and aging. Proc Natl Acad Sci USA 103:12517–12522

Nishida Y, Yokota T, Takahashi T, Uchihara T, Jishage K, Mizusawa H 2006 Deletion of vitamin E enhances phenotype of Alzheimer disease model mouse. Biochim Biophys Res Commun 350:530–536

Panov AV, Gutekunst CA, Leavitt BR et al 2002 Early mitochondrial calcium defects in Huntington's disease are a direct effect of polyglutamines. Nat Neurosci 5:731–736

Pasinelli P, Belford ME, Lennon N et al 2004 Amyotrophic lateral sclerosis-associated SOD1 mutant proteins bind and aggregate with Bcl-2 in spinal cord mitochondria. Neuron 43:19–30

Pastorino L, Sun A, Lu PJ et al 2006 The prolyl isomerase Pin1 regulates amyloid precursor protein processing and amyloid-beta production. Nature 440:528–534

Pratico D, Uryu K, Leight S, Trojanowski JQ, Lee VM 2001 Increased lipid peroxidation precedes amyloid plaque formation in an animal model of Alzheimer amyloidosis. J Neurosci 21:4183–4187

Pyle A, Foltynie T, Tiangyou W et al 2005 Mitochondrial DNA haplogroup cluster UKJT reduces the risk of PD. Ann Neurol 57:564–567

St-Pierre J, Drori S, Uldry M et al 2006 Suppression of reactive oxygen species and neurodegeneration by the PGC-1 transcriptional coactivators. Cell 127:397–408

Trifunovic A 2006 Mitochondrial DNA and ageing. Biochim Biophys Acta 1757:611–617

Valente EM, Salvi S, Ialongo T et al 2004 PINK1 mutations are associated with sporadic early-onset parkinsonism. Ann Neurol. 56:336–341

Vijayvergiya C, Beal MF, Buck J, Manfredi G 2005 A portion of mutant hSOD1 localizes on the matrix side of the inner mitochondrial membrane and forms aberrant macromolecular aggregates in fALS mice. J Neurosci 25:2463–2470

Wang D, Qian L, Xiong H et al 2006 Antioxidants protect PINK1-dependent dopaminergic neurons in Drosophila. Proc Natl Acad Sci USA. 103:13520–13525

Weydt P, Pineda VV, Torrence AE et al 2006 Thermoregulatory and metabolic defects in Huntington's disease transgenic mice implicate PGC-1 alpha in Huntington's Disease neurodegeneration. Cell Metab 4:349–362

Whitworth AJ, Theodore DA, Greene JC, Benes H, Wes PD, Pallanck LJ 2005 Increased glutathione S-transferase activity rescues dopaminergic neuron loss in a Drosophila model of Parkinson's Disease. Proc Natl Acad Sci USA 102:8024–8029

Yang Y, Gehrke S, Imai Y et al 2006 Mitochondrial pathology and muscle and dopaminergic neuron degeneration caused by inactivation of Drosophila Pink1 is rescued by Parkin. Proc Natl Acad Sci USA 103:10793–10798

DISCUSSION

Duchen: You have covered so much it is hard to know where to start. One of the key issues seems to be the nature of the tissue or cell selectivity of the problem. For example, you have SOD mutations in every tissue in the ALS mouse model and yet it is only the motor neurons that degenerate. How could one account for that?

Beal: I will stay with ALS as an example, but we could go through each disease in the same way. In ALS there is less of an association of mutant SOD1 in the liver mitochondria than there is the brain. We don't know why. Also, the spinal cord mitochondria seem to be more vulnerable than the others. Motor neurons have the longest axons in the body. The cell body is 1% of the neuron, and the long axoplasm has to be maintained. This is one hypothesis: this may make them more vulnerable.

Larsson: In one of your models, the SOD2 knockout, you observe more plaques in the heterozygous knockout. There are indications that plaque formation may be protective against neurodegeneration. This has been reported in cellular models of HD. Sequestration of these toxic proteins into plaques could be a good thing. Perhaps we should try to increase oxidative stress in patients in order to accomplish this!

Spiegelman: I think they did that experiment in the Vietnam war. It didn't turn out well!

Beal: Your point is well taken, and in HD this is unequivocal. Most people think these plaques may just be a marker for what else is going on, and it is the small oligomers which are critical to pathogenesis.

Larsson: I heard a neuropathologist say that it is impossible to diagnose Alzheimer's by just slicing a brain, because plaques and other changes are also seen in normal human ageing. It is the clinical picture together with the neuropathology that is critical.

Beal: Yes, they occur in normal ageing and some of those people don't seem to be demented. But overall there is a good correlation between plaques and tangles and dementia. The tangles correlate better than the plaques.

Spiegelman: I'd like to ask about PGC1α and HD. First of all, the two papers that were in *Cell* and *Cell Metabolism* weren't particularly consistent with each other. One paper shows no defect in transcription of PGC1α. The mechanisms they suggest are quite different.

Regarding the brown fat, first of all mice are more dependent on brown fat thermogenesis when they are younger than they are when they get older. If the change in temperature is age related, it is going in the wrong direction for being controlled by brown fat. Young mice and humans are dependent on brown fat; when you get bigger it becomes less important because of surface area:volume ratios, hair and shivering. It has been shown unambiguously that there are morphological changes in brown fat. The problem is that brown fat is driven by sympathetic innervation: just because there is a brown fat defect, it can't be assumed that it is cell autonomous. Someone needs to inject a β3 adrenergic agonist into those animals, getting the nervous system out of the way: although you have a whole animal, the agonist makes it tissue autonomous. Then you ask what the response is like. If it is dysregulated you have taken the nervous system out of the picture. You can then probably ascribe it to a tissue-autonomous defect.

Nicholls: Have there been any classical restoration studies with brown adipocytes in these animals under these conditions?

Spiegelman: I don't think so. The trouble with the interpretation of all those experiments is that the brown fat is terribly dependent on sympathetic activity. If you see a defect in brown fat you can't ascribe it to the brown fat itself. It could well be CNS centred, even though your readout is brown fat. You need to inject a β3 agonist and take the sympathetic nervous system out of the equation. If you have a UCP gene defect, then it is the simplest experiment. You come back in two or three hours and look at the induction of that gene. What the nervous system is doing becomes irrelevant.

Nicholls: It is pertinent that these effects are showing up in the ageing animal. Depending on the temperature in which mice have been reared, temperature may have had little effect.

Spiegelman: It is not implausible that it is a brown fat defect, but the fact that the defect emerges as the animal gets older makes it look less like a brown fact defect.

Turnbull: What is the level of exercise of these mice? Are they pursuing the same level of activity? This would have a secondary effect.

Beal: They clearly have reduced activity as they get older. If you exercise them more and put them in an enriched environment they live longer. One interesting experiment that may be pertinent to the temperature issue is that if they put the R62 mice into elevated temperature (32 °C) they lived longer.

Spiegelman: I don't know how to interpret that.

Nicholls: Is it possible that there is a central defect in temperature homeostasis in the ventromedial hypothalamus?

Spiegelman: That is what occurred to us. This needs a tissue or cell autonomous assay. There is so much going on, and the CNS is known to be screwed up. We need a molecular defect that we can get a handle on.

Nicholls: Some years ago we and others did some work with either the isolated mitochondria or the isolated brown adipocytes, just checking the functionality. Maintaining mice at different temperatures is critical, because the brown fat is cold induced at any stage in a small rodent. Is there a loss of cold-induced thermogenesis in these animals?

Spiegelman: That could be a CNS defect.

Beal: Your PGC1α knockout mice show exactly the same defects.

Spiegelman: There is a strong defect when they are young, and this lessens as they get older. This is what you'd expect if it is brown fat: they are less dependent on it as they get bigger and hairier.

Nicholls: So PGC1α is a global transcription regulator in all tissues. Why does it focus on brown fat?

Spiegelman: If you do a tissue blot for PGC1α or β, and a blot for mitochondrial DNA, there is a very good correlation. Tissues with lots of mitochondria have it, including brown fat. It is already high and it is cold inducible. Its core job is mitochondrial biogenesis.

Turnbull: Regarding the data on the control region, I have never understood how changes in the control region could affect Alzheimer's disease. This has eluded me. And we haven't been able to replicate the data.

Beal: This is Doug Wallace's work. He says that it potentially will reduce the overall transcription of mitochondrial DNA.

Turnbull: All the evidence is suggesting that many of the changes we are seeing in Alzheimer's disease are increasing oxidative damage. It seems perverse to me that you are going to decrease any ROS produced by this sort of mechanism.

Beal: It would not necessarily reduce ROS.

Spiegelman: Having messed up mitochondria could conceivably make it worse.

Jacobs: In any case, there is no direct evidence that these mutations do influence DNA replication and transcription.

Larsson: There are many reports of mutations in control regions that affect binding sites for mitochondrial transcription factor A. We need to test these mutations in an *in vitro* transcription system.

Schon: It might be irrelevant anyway, if it is heteroplasmic and it is at 2%.

Larsson: I agree that it might not impact overall mtDNA transcription if there is a heteroplasmic situation with low levels of the mutation, however, *in vitro* experiments will clarify whether these mutations have the capacity to impair mtDNA transcription.

Rizzuto: I was particularly interested in one bit of data I was unaware of, which is the idea that in Parkinson's disease there is a selective increase in mtDNA deletions. Do you have a hint of why it happens there? Is it the rate of the generation, or could it be autophagic clearance of the wrong mitochondria?

Turnbull: I agree that this is a surprising observation. Quite clearly, it is cell-type specific, and it is a substantia nigra neuron. It isn't present as much in other CNS neurons. It is thought that these neurons have a very high level of oxidative damage, so it could be damage or replication errors or repair. These are clonally expanded mutations. Each neuron has a different deletion and this deletion is clonally expanded to high levels. Whether or not it is impaired replication, excess repair, or impaired autophagic clearance is unclear. This is one of the things we are interested in finding out.

Rizzuto: Apart from the intrinsic interest for the disease and the substantia nigra degeneration, it is important for us to understand the mechanism of mitochondrial turnover and clearance.

Turnbull: Absolutely. It is difficult to study in these neurons because we really don't know the rate of replication.

Schon: Kearns-Sayre syndrome, which I'll mention in my paper, is a deletion disorder. Patients with it have the deletion in all cells of the body, yet if you look in the brain, for example, there is a predilection for the deletion to accumulate remarkably in the choroid plexus. This doesn't happen with other mutations. Even though the choroid plexus is the progeny of ependymal cells, so there might be cell turnover, it is hard to imagine that it is merely turnover by cells because then you would see this in just about every mitochondrial disorder. The cell specificity is a conundrum. In fact, if you had deletions in Parkinson's disease patients, one might ask whether there are deletions in choroid plexus.

Turnbull: Don't get me wrong: we don't just see spontaneous deletions in substantia nigra neurons; they are clearly present in a lot of post-mitotic cells. It was the high level of deletions which was completely unexpected, because in muscle or other cell types the level is so much lower.

Jacobs: One thing we need to know is the rate of turnover of mtDNA, which may then be a secondary consequence of an increased amount of DNA damage and the system turning over, to get rid of the damaged DNA. The deleted molecules could arise as a by-product of that, rather than being directly caused by DNA-damaging agents.

Turnbull: I agree. If we could work out the mechanism for measuring the turnover, it would be helpful.

Schon: This has bothered me for a long time. Many people talk about the creation of deletions as somehow being the product of free radical damage. To me it is a complete mystery how you can generate a deletion merely through this.

Nicholls: What is the mechanism of deletion?

Schon: Half of all deletions are mediated by short direct repeats. It is a replication or recombination mediated event in which there are dissimilar pieces of DNA joined together. To invoke point mutations that are generating free radicals is special pleading.

Jacobs: The prediction is that you would see a higher rate of point mutations in molecules that have deletions, and this isn't the case.

Mitochondrial dysfunction in mammalian ageing

Mügen Terzioglu and Nils-Göran Larsson[1]

Department of Laboratory Medicine, Karolinska Institutet, S-14186 Stockholm, Sweden

Abstract. Ageing is likely a multifactorial process caused by accumulated damage to a variety of cellular components. Increasing age in mammals correlates with increased levels of mitochondrial DNA (mtDNA) mutations and deteriorating respiratory chain function. Mosaic respiratory chain deficiency in a subset of cells in various tissues, such as heart, skeletal muscle, colonic crypts and neurons, is typically found in aged humans. Experimental evidence in the mouse has linked increased levels of somatic mtDNA mutations to a variety of ageing phenotypes, such as osteoporosis, hair loss, greying of the hair, weight reduction and decreased fertility. It has been known for a long time that respiratory chain-deficient cells are more prone to undergo apoptosis and increased cell loss is therefore likely of importance in age-associated mitochondrial dysfunction. There is a tendency to automatically link mitochondrial dysfunction to increased production of reactive oxygen species (ROS). However, the experimental support for this concept is rather weak. Mouse models with respiratory chain deficiency induced by tissue-specific mtDNA depletion or by massive increase of point mutations in mtDNA have very minor or no increase of oxidative stress. Future studies are needed to address the relative importance of mitochondrial dysfunction and ROS in mammalian ageing.

2007 Mitochondrial biology: new perspectives. Wiley, Chichester (Novartis Foundation Symposium 287) p 197–213

Ageing can be defined as '*a progressive, generalized impairment of function, resulting in an increased vulnerability to environmental challenge and a growing risk of disease and death*' (Kirkwood 2005). It is generally assumed that accumulated damage to a variety of cellular systems is the underlying cause of ageing (Kirkwood 2005). It should be noted that ageing is frequently not observed in nature as most wild animal species succumb to a variety of environmental insults (e.g. cold, starvation, predation, accidents) before they reach old age (Fig. 1). It has been argued that there has been little evolutionary pressure to maintain efficient repair and maintenance mechanisms at high ages as most animals do not reach high ages in the

---

[1]This paper was presented at the symposium by Nils-Göran Larsson to whom correspondence should be addressed.

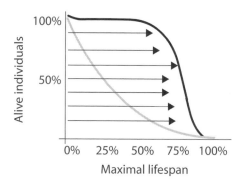

FIG. 1. Schematic illustration showing mechanisms causing ageing (adapted from Kirkwood 2005). Ageing results from accumulated damage to multiple cellular systems in combination with failure of essential maintenance and repair systems (arrows), e.g. ROS scavenging enzymes, DNA repair enzymes, protein degradation machineries. Ageing (black curve) is only seen in protected environments. Most animals in the wild die (grey curve) because of external causes, e.g. cold, starvation, predation and accidents. Evolution has designed the maintenance and repair systems (arrows) to function efficiently during the expected lifespan of the individual in the wild (grey line).

wild (Kirkwood 2005). Ageing is therefore mainly observed in animal populations that live in protected environments, e.g. in a mouse house. The disposable soma theory predicts that stochastic accumulated damage and decline of somatic maintenance and repair systems cause ageing in somatic tissues (Fig. 1), whereas the germline is carefully maintained and kept immortal (Kirkwood 2005). The robustness of the maintenance and repair mechanisms will of course be under genetic control, but there are strong arguments that the decay associated with ageing is not genetically programmed (Austad 2004, Kirkwood 2005).

Caloric restriction prolongs life

Caloric restriction (CR), also referred to as dietary restriction, prolongs lifespan in a variety of experimental organisms including budding yeast, worms, flies and rodents (Bordone & Guarente 2005, Weindruch & Walford 1988). It is currently not clear if CR can also extend lifespan in non-human primates and humans. CR reduces the incidence of age-associated diseases such as cancer, cardiovascular diseases and diabetes in mammals (Bordone & Guarente 2005, Weindruch & Walford 1988). There is often a reverse relationship between fecundity and lifespan so that organisms with long spans tend to have low fecundity and *vice versa* (Partridge et al 2005). CR causes low fecundity and it has been speculated that an organism that invests less in reproduction may have more resources to allocate for maintenance functions. The CR seems to be conserved during animal evolution and there is intense ongoing research to define the underlying molecular pathways (Guarente & Picard 2005). Nitric oxide (NO) has been shown to have the capacity to induce mitochondrial biogenesis in mice (Nisoli et al 2003). Interestingly, CR increases NO production by inducing the expression of endothelial NO synthetase and thereby causes increased mitochondrial biogenesis (Nisoli et al 2005).

Single gene mutants can extend lifespan in experimental animals

Pioneering genetic experiments in the worm *C. elegans* showed that it is possible to perform genetic screens to obtain long-lived mutants (Klass 1983). Subsequent work led to the identification of the *age-1* gene (Friedman & Johnson 1988). Genes whose loss of function extend lifespan in the worm can be divided into three main groups affecting: (i) dauer formation (daf genes and additional genes), (ii) physiological rates (e.g. Clk genes) and (iii) normal feeding (eat genes) (Feng et al 2001). Genes that influence dauer formation include *daf-2* and *daf-16*, which encode components of the insulin/IGF1 signalling pathway (Feng et al 2001). Interestingly, mutations decreasing the insulin/IGF1 signalling can prolong the lifespan of worms, flies and mice. Also mutations in the mouse that affect the growth hormone axis and thereby secondarily influence the insulin/IGF1 signalling prolong lifespan in the mouse (Liang et al 2003). Mitochondrial dysfunction in worms can increase their lifespan (Feng et al 2001) and a recent report suggests that this may also be the case in the mouse (Dell'agnello et al 2007).

The respiratory chain

The mitochondrial network is the main producer of cellular ATP, the energy currency used for a variety of metabolic reactions (Fig. 2). Degradation of nutrients absorbed from food, e.g. glucose and fatty acids, transfers electrons to carrier molecules such as NAD^+ and FAD^+ thus generating NADH and $FADH_2$, which, in turn, deliver electrons to the respiratory chain located in the inner mitochondrial membrane (Saraste 1999). The oxidative phosphorylation system consists of five enzyme complexes (Fig. 2). Complexes I-IV constitute the electron transport chain that uses the electron transfer process to provide energy for proton translocation. In the final step of this process the electrons are reducing O_2 to form water. The ATP-synthase, sometimes also referred to as complex V, uses the proton gradient to generate ATP by a rotary mechanism. There is normally a tight coupling between the electron transport and ATP synthesis so that inhibition of ATP synthase will inhibit the electron transport and cellular respiration (Cannon et al 2006). Protonophores, e.g. FCCP, and the specialized protein, uncoupling protein 1, can uncouple electron transport from ATP synthesis and thereby cause low ATP production despite high cellular respiration (Cannon et al 2006).

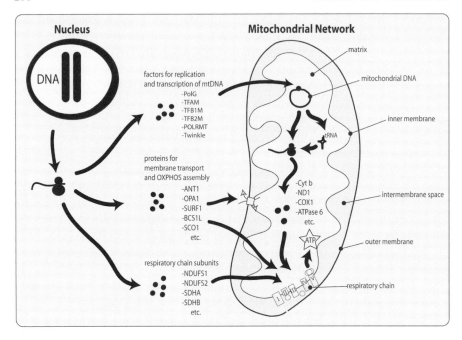

FIG. 2. The production of cellular ATP in the mitochondrial network. Mitochondrial biogenesis and function are under dual genetic control requiring both nuclear and mitochondrial DNA (mtDNA) genes. This cross-talk between the two genomes is essential for maintaining the structure and function of mitochondria. The mtDNA encodes RNA components of the mitochondrial translation apparatus and 13 proteins, all of which are essential for the function of the respiratory chain. Nuclear genes encode all other mitochondrial proteins, which include the majority of the respiratory chain subunits (e.g. NDUFS1, NDUFS2, SDHA, etc.), all proteins necessary for transcription and replication of mtDNA (e.g. PolG, TFAM, TFB1M etc.) and all proteins necessary for transport across the mitochondrial membranes and assembly of the respiratory chain (e.g. ANT1, OPA1, SURF1, etc.).

Genetics of mitochondrial DNA (mtDNA)

The mitochondria in eukaryotes constitute a dynamic network that is continuously fusing and dividing (Yaffe 2003). Interestingly, reducing mitochondrial fission can result in a longer lifespan in budding yeast (Scheckhuber et al 2007). There are 10^3–10^4 copies of mtDNA in a typical somatic mammalian cell (Larsson & Clayton 1995). The mtDNA encodes 13 proteins, all of which are constituents of the respiratory chain, and 22 tRNAs and 2 rRNAs, which are needed for the mitochondrial proteins synthesis (Fig. 2). However, the remaining ~10^3 different mitochondrial proteins are encoded by nuclear genes, including all proteins necessary for transcription and replication of mtDNA and the majority of the proteins of the respiratory chain (Fig. 2). The high copy number of mtDNA in mammalian cells will ensure that mutations affecting a single copy will not impact overall mitochondrial function. In fact, it was long assumed that the high copy number would be protective and prevent expansion of *de novo* mutations arising in a single mtDNA molecule. It therefore came as a big surprise when two independent groups in 1988 found pathogenic mtDNA mutations in human patients with mitochondrial diseases (Holt et al 1988, Wallace et al 1988). The affected patients either had homoplasmy, i.e. only mutated mtDNA (Wallace et al 1988), or heteroplasmy, i.e. a mixture of wild-type and mutated mtDNA (Holt et al 1988). Heteroplasmic mtDNA mutations are subject to mitotic segregation so that they can accumulate to very high levels in some cells, thus creating a mosaic pattern of respiratory chain deficiency in affected tissues (Fig. 3). The reason for this phenomenon is that the replication of mtDNA occurs independently of the cell cycle and that even postmitotic cells continuously replicate their mtDNA (Fig. 3).

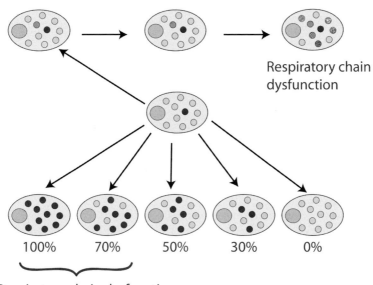

FIG. 3. Clonal expansion of mtDNA mutations. There are 10^3–10^4 copies of mtDNA in a somatic mammalian cell. Multiple *de novo* mutations of mtDNA affecting a single cell may cause respiratory chain dysfunction (upper row). There is no coupling between the cell cycle and replication of mtDNA and therefore a somatic mtDNA mutation affecting a single mtDNA molecule may undergo clonal expansion (lower row). Such clonal expansion may also occur in postmitotic cells. Clonal expansion of an mtDNA molecule containing a pathogenic mutation will create a mosaic pattern of respiratory chain deficiency.

A bottleneck for mtDNA transmission

The existence of a bottleneck for mtDNA transmission during oogenesis was first reported in Holstein cows (Lapis et al 1988) and has since been confirmed by numerous studies of human pedigrees (Larsson et al 1992, Shoubridge & Wai 2007). Pioneering experimental work in this area has shown that mice engineered to harbour neutral heteroplasmic mtDNA mutations demonstrate a bottleneck phenomenon during mtDNA transmission and that it occurs early in oogenesis (Jenuth et al 1996). The conservation of the bottleneck mechanism among mammals points towards an important evolutionary role and additional work is warranted to elucidate the underlying molecular mechanisms.

Theoretical considerations have suggested that the high mutation rate, the maternal transmission and the absence of biparental mtDNA recombination should leave mtDNA vulnerable to Müller's ratchet, a process whereby deleterious mutations accumulate in asexual lineages, leading to mutational meltdown of the genome (Elson & Lightowlers 2006). The bottleneck mechanism may decrease the risk of germline transmission of mtDNA mutations and may thereby prevent or at least slow down the Müller's ratchet process (Elson & Lightowlers 2006). Additional, yet unidentified, mechanisms may also have roles in decreasing transmission of mutated mtDNA in maternal lineages.

Increased levels of somatic mtDNA mutations in ageing

In 1988, Pikó studied the integrity of mtDNA in young adult and senescent rats by performing electron microscopic studies of reconstituted mtDNA duplexes and reported an increased abundance of structural aberrations consistent with mtDNA deletions in aged animals (Pikó et al 1988). Shortly thereafter reports started to appear that rearrangements of mtDNA could be identified by PCR analyses of tissues from aged humans (Corral-Debrinski et al 1992, Sato et al 1989, Soong et al 1992). It was also reported that the respiratory chain capacity decreases with age in various tissues such as skeletal muscle (Trounce et al 1989) and liver (Yen et al 1989). Hypotheses were put forward that acquired mutations of mtDNA would increase with time and segregate in mitotic tissues to eventually cause decline of respiratory chain function leading to age-associated degenerative disease and ageing (Linnane et al 1989).

Sequencing of mtDNA in fibroblasts from young and aged humans has demonstrated the occurrence of age-associated, site-specific mutations in the larger non-coding control region (Michikawa et al 1999). Some of these mutations alter highly conserved sequences necessary for transcription, however, experimental studies demonstrating any functional importance of these mutations are still lacking.

The data obtained from studies of tissues from humans and other mammals have provided circumstantial evidence for the involvement of mtDNA mutations in ageing (Cottrell & Turnbull 2000). Recent experimental work has provided clear evidence that somatic mtDNA mutations indeed can cause a variety of ageing phenotypes such as osteoporosis, hair loss, alopecia, leanness, weight loss, reduced fertility and heart enlargement (Trifunovic et al 2004). The mtDNA mutator mouse (Fig. 4) is engineered to have a deficient proof-reading activity of the mitochondrial DNA polymerase and as a consequence it develops high levels of a random set of point mutations in mtDNA (Trifunovic et al 2004). It is important to point out that this mouse model shows that point mutations of mtDNA can cause ageing phenotypes if present at high enough levels. However, this mtDNA mutator mouse experiment of course does not prove that the levels present in normal ageing are sufficient to cause ageing phenotypes. Further experiments are clearly needed in this area.

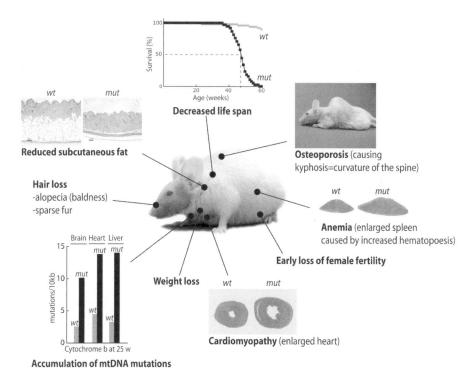

FIG. 4. The mtDNA mutator mouse displays signs of premature ageing. The mtDNA mutator mouse is homozygous for a knock-in mutation and expresses a proof reading-deficient mtDNA polymerase. This causes the accumulation of somatic point mutations of mtDNA as well as low levels of deleted mtDNA. The mouse develops a progressive respiratory chain deficiency and shortened lifespan accompanied by a variety of ageing phenotypes.

Focal respiratory chain deficiency in ageing

In 1989, Müller-Höcker published a carefully conducted extensive enzyme histochemical study of hearts from individuals of different ages. Surprisingly, he found that increasing age was associated with focal respiratory chain deficiency in a subset of the cardiomyocytes (Müller-Höcker 1989). Respiratory chain deficient cells could be seen sporadically from age 20 years and onwards and were present in all individuals above the age of 60 years (Müller-Höcker 1989). Müller-Höcker also studied skeletal muscle specimens and found a similar pattern of age-associated focal respiratory chain dysfunction (Müller-Höcker 1990). Subsequent studies have confirmed these findings and demonstrated that focal respiratory chain deficiency is often associated with clonal expansion of deleted mtDNA mutations (Fig. 3). Besides heart and skeletal muscle, age-associated focal respiratory chain deficiency has also been reported in colonic crypts (Taylor et al 2003) and a variety of neuronal cell types, such as hippocampal neurons and midbrain dopaminergic neurons (Cottrell et al 2001, 2002). The idea that somatic mtDNA mutations may cause Parkinson's disease has been around for a long time. Reports in the early 1990s showed increased levels of mtDNA deletions in striatum (Ikebe et al 1990) and semi-quantitative PCR analyses of different brain regions showed a region-specific variation in the mutation load with high mutation levels in substantia nigra and the basal ganglia (Corral-Debrinski et al 1992, Soong et al 1992). A recent study used an *in situ* approach to analyse mutation levels in individual dopamine neurons and found high levels of mtDNA deletions (Bender et al 2006). This mutational load is likely of functional significance as aged humans and Parkinson's disease patients have an increase frequency of respiratory chain deficient dopamine neurons (Bender et al 2006). Genetic disruption of mtDNA expression in dopamine neurons of the mouse reproduces many, but not all, key features of Parkinson's disease, e.g. slowly progressive locomotion impairment with a typical denervation pattern of striatum and dopamine nerve cell loss accompanied by inclusion formation (Ekstrand et al 2007).

Does mitochondrial dysfunction cause increased oxidative damage?

There is an unfortunate tendency to automatically associate respiratory chain dysfunction with increased ROS production. The experimental support for this link is weak and there are in fact many examples of mouse models with impaired mitochondrial function that only exhibit minor or no oxidative stress (Kujoth et al 2005, Trifunovic et al 2005, Wang et al 2001). It is important to point out that oxidative damage will only occur if ROS production exceeds the capacity of the defence mechanisms inactivating the different ROS species (Fig. 5). Furthermore, the steady-state levels of oxidatively damaged molecules depend on both

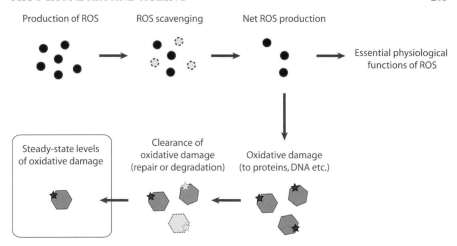

FIG. 5. Schematic representation of reactive oxygen species (ROS) formation and oxidative damage. ROS production is counteracted by a variety of ROS defence mechanisms resulting in net ROS production. ROS may have a function in normal cell physiology and also cause damage to a variety of macromolecules such as DNA, lipids and proteins. The steady-state levels of oxidatively damaged molecules are determined by formation and clearance.

net ROS formation and clearance of damaged molecules (Fig. 5). It is even, in principle, possible that increased levels of oxidatively damaged molecules may be present even if ROS production is normal, i.e. if the clearance of ROS-damaged molecules is defective.

Mitochondria are assumed to be the main cellular producers of ROS, consistent with the fact that mitochondrial enzymes transport electrons and consume most of the cellular oxygen in the process of oxidative phosphorylation (Orrenius et al 2007). It is estimated that ~0.2% of the oxygen used in cellular respiration is converted to ROS (Balaban et al 2005). Mouse knockout studies have demonstrated that ROS production in the mitochondrial matrix is scavenged by a superoxide dismutase (SOD2 or MnSOD) and inactivation of this scavenging enzyme leads to grave pathology and premature death (Melov et al 1998). Many studies point to complex I and complex III as critical sites for generation of mitochondrial ROS (Balaban et al 2005). However, it should be emphasized that although several mitochondrial enzymes have the capacity to produce ROS *in vitro* (Andreyev et al 2005), there is no general consensus on the *in vivo* role for most of these sites.

Mitochondrial dysfunction induces apoptosis

Studies in the 1950s and 1960s clearly implicated mitochondrial dysfunction in radiation-induced pyknotic cell death (Orrenius et al 2007). The identification of apoptosis as a specific genetically programmed form of cell death and the finding that cytochrome c is released by mitochondria during apoptosis soon stimulated intense research on the role of apoptosis in human pathology (Orrenius et al 2007). It was initially reported that cancer cell lines lacking mtDNA had a normal execution of apoptosis and even showed resistance to apoptosis induction (Jacobson et al 1993). However, subsequent studies demonstrated that cell lines and animal tissues with reduced respiratory chain function actually were more prone to undergo apoptosis (Wang et al 2001). Mouse embryos with homozygous germ line disruption of the *Tfam* gene develop massive apoptosis and die in midgestation (Wang et al 2001). Respiratory chain dysfunction in differentiated cell types such as cardiomyocytes and pyramidal neurons also causes increased cell death consistent with apoptosis (Sorensen et al 2001). Apoptosis is also a prominent feature in the mtDNA mutator mouse (Kujoth et al 2005).

Acknowledgements

This study was supported by the Swedish Research Council, the Swedish Heart and Lung Foundation, the Torsten and Ragnar Söderbergs Foundation, The Swedish Strategic Foundation (INGVAR) and Knut and Alice Wallenbergs Stiftelse.

References

Andreyev AY, Kushnareva YE, Starkov AA 2005 Mitochondrial metabolism of reactive oxygen species. Biochemistry (Mosc) 70:200–214
Austad SN 2004 Is aging programed? Aging Cell 3:249–251
Balaban RS, Nemoto S, Finkel T 2005 Mitochondria, oxidants, and aging. Cell 120:483–495
Bender A, Krishnan KJ, Morris CM et al 2006 High levels of mitochondrial DNA deletions in substantia nigra neurons in aging and Parkinson disease. Nat Genet 38:515–517
Bordone L, Guarente L 2005 Calorie restriction, SIRT1 and metabolism: understanding longevity. Nat Rev Mol Cell Biol 6:298–305
Cannon B, Shabalina IG, Kramarova TV, Petrovic N, Nedergaard J 2006 Uncoupling proteins: a role in protection against reactive oxygen species—or not? Biochim Biophys Acta 1757:449–458
Corral-Debrinski M, Horton T, Lott MT, Shoffner JM, Beal MF, Wallace DC 1992 Mitochondrial DNA deletions in human brain: regional variability and increase with advanced age. Nat Genet 2:324–329
Cottrell DA, Turnbull DM 2000 Mitochondria and ageing. Curr Opin Clin Nutr Metab Care 3:473–478
Cottrell DA, Blakely EL, Johnson MA, Ince PG, Borthwick GM, Turnbull DM 2001 Cytochrome c oxidase deficient cells accumulate in the hippocampus and choroid plexus with age. Neurobiol Aging 22:265–272
Cottrell DA, Borthwick GM, Johnson MA, Ince PG, Turnbull DM 2002 The role of cytochrome c oxidase deficient hippocampal neurones in Alzheimer's disease. Neuropathol Appl Neurobiol 28:390–396
Dell'agnello C, Leo S, Agostino A et al 2007 Increased longevity and refractoriness to Ca(2+)-dependent neurodegeneration in Surf1 knockout mice. Hum Mol Genet 16:431–444
Ekstrand MI, Terzioglu M, Galter D et al 2007 Progressive parkinsonism in mice with respiratory-chain-deficient dopamine neurons. Proc Natl Acad Sci USA 104:1325–1330

Elson JL, Lightowlers RN 2006 Mitochondrial DNA clonality in the dock: can surveillance swing the case? Trends Genet 22:603–607

Feng J, Bussiere F, Hekimi S 2001 Mitochondrial electron transport is a key determinant of life span in Caenorhabditis elegans. Dev Cell 1:633–644

Friedman DB, Johnson TE 1988 Three mutants that extend both mean and maximum life span of the nematode, Caenorhabditis elegans, define the age-1 gene. J Gerontol 43:B102–B109

Guarente L, Picard F 2005 Calorie restriction–the SIR2 connection. Cell 120:473–482

Holt IJ, Harding AE, Morgan-Hughes JA 1988 Deletions of muscle mitochondrial DNA in patients with mitochondrial myopathies. Nature 331:717–719

Ikebe S-I, Tanaka M, Ohno K et al 1990 Increase of deleted mitochondrial DNA in the striatum in Parkinson's disease and senescence. Biochem Biophys Res Commun 170:1044–1048

Jacobson MD, Burne JF, King MP, Miyashita T, Reed JC, Raff MC 1993 Bcl-2 blocks apoptosis in cells lacking mitochondrial DNA. Nature 361:365–369

Jenuth JP, Peterson AC, Fu K, Shoubridge EA 1996 Random genetic drift in the female germline explains the rapid segregation of mammalian mitochondrial DNA. Nat Genet 14:146–151

Kirkwood TB 2005 Understanding the odd science of aging. Cell 120:437–447

Klass MR 1983 A method for the isolation of longevity mutants in the nematode Caenorhabditis elegans and initial results. Mech Ageing Dev 22:279–286

Kujoth GC, Hiona A, Pugh TD et al 2005 Mitochondrial DNA mutations, oxidative stress, and apoptosis in mammalian aging. Science 309:481–484

Lapis PJ, VanDeWalle MJ, Hauswirth WW 1988 Unequal partitioning of bovine mitochondrial genotypes among siblings. Proc Natl Acad Sci USA 85:8107–8110

Larsson NG, Clayton DA 1995 Molecular genetic aspects of human mitochondrial disorders. Annu Rev Genet 29:151–178

Larsson NG, Tulinius MH, Holme E et al 1992 Segregation and manifestations of the mtDNA tRNALys A->G(8344) mutation of myoclonus epilepsy and ragged-red fibers (MERRF) syndrome. Am J Hum Genet 51:1201–1212

Liang H, Masoro EJ, Nelson JF, Strong R, McMahan CA, Richardson A 2003 Genetic mouse models of extended lifespan. Exp Gerontol 38:1353–1364

Linnane AW, Marzuki S, Ozawa T, Tanaka M 1989 Mitochondrial DNA mutations as an important contributor to ageing and degenerative diseases. Lancet 1:642–645

Melov S, Schneider JA, Day BJ et al 1998 A novel neurological phenotype in mice lacking mitochondrial manganese superoxide dismutase. Nat Genet 18:159–163

Michikawa Y, Mazzucchelli F, Bresolin N, Scarlato G, Attardi G 1999 Aging-dependent large accumulation of point mutations in the human mtDNA control region for replication. Science 286:774–779

Müller-Höcker J 1989 Cytochrome-c-oxidase deficient cardiomyocytes in the human heart— an age-related phenomenon. Am J Pathol 134:1167–1173

Müller-Höcker J 1990 Cytochrome c oxidase deficient fibres in the limb muscle and diaphragm of man without muscular disease: an age related alteration. J Neurol Sci 100:14–21

Nisoli E, Clementi E, Paolucci C et al 2003 Mitochondrial biogenesis in mammals: the role of endogenous nitric oxide. Science 299:896–899

Nisoli E, Tonello C, Cardile A et al 2005 Caloric restriction promotes mitochondrial biogenesis by inducing the expression of eNOS. Science 310:314–317

Orrenius S, Gogvadze V, Zhivotovsky B 2007 Mitochondrial oxidative stress: implications for cell death. Annu Rev Pharmacol Toxicol 47:143–183

Partridge L, Gems D, Withers DJ 2005 Sex and death: what is the connection? Cell 120:461–472

Pikó L, Hougham AJ, Bulbitt KJ 1988 Studies of sequence heterogeneity of mitochondrial DNA from rat and mouse tissues: evidence for an increased frequency of deletions/additions with ageing. Mech Ageing Dev 43:279–293

Saraste M 1999 Oxidative phosphorylation at the fin de siecle. Science 283:1488–1493
Sato W, Tanaka M, Ohno K, Yamamoto T, Takada G, Ozawa T 1989 Multiple populations of deleted mitochondrial DNA detected by a novel gene amplification method. Biochem Biophys Res Commun 162:664–672
Scheckhuber CQ, Erjavec N, Tinazli A, Hamann A, Nystrom T, Osiewacz HD 2007 Reducing mitochondrial fission results in increased life span and fitness of two fungal ageing models. Nat Cell Biol 9:99–105
Shoubridge EA, Wai T 2007 Mitochondrial DNA and the mammalian oocyte. Curr Top Dev Biol 77:87–111
Soong NW, Hinton DR, Cortopassi G, Arnheim N 1992 Mosaicism for a specific somatic mitochondrial DNA mutation in adult human brain. Nat Genet 2:318–323
Sorensen L, Ekstrand M, Silva JP et al 2001 Late-onset corticohippocampal neurodepletion attributable to catastrophic failure of oxidative phosphorylation in MILON mice. J Neurosci 21:8082–8090
Taylor RW, Barron MJ, Borthwick GM et al 2003 Mitochondrial DNA mutations in human colonic crypt stem cells. J Clin Invest 112:1351–1360
Trifunovic A, Wredenberg A, Falkenberg M et al 2004 Premature ageing in mice expressing defective mitochondrial DNA polymerase. Nature 429:417–423
Trifunovic A, Hansson A, Wredenberg A et al 2005 Somatic mtDNA mutations cause aging phenotypes without affecting reactive oxygen species production. Proc Natl Acad Sci USA 102:17993–17998
Trounce I, Byrne E, Marzuki S 1989 Decline in skeletal muscle mitochondrial respiratory chain function: possible factor in ageing. Lancet 1:637–639
Wallace DC, Singh G, Lott MT et al 1988 Mitochondrial DNA mutation associated with Leber's hereditary optic neuropathy. Science 242:1427–1430
Wang J, Silva J, Gustafsson CM, Rustin P, Larsson NG 2001 Increased in vivo apoptosis in cells lacking mitochondrial DNA gene expression. Proc Natl Acad Sci USA 98:4038–4043
Weindruch RW, Walford RL 1988 The retardation of aging and disease by dietary restriction. Thomas, Springfield IL
Yaffe MP 2003 The cutting edge of mitochondrial fusion. Nat Cell Biol 5:497–499
Yen T-C, Chen Y-S, King K-L, Yeh S-H, Wei Y-H 1989 Liver mitochondrial respiratory functions decline with age. Biochem Biophys Res Commun 165:994–1003

DISCUSSION

Nicholls: In these induced mitochondrial deficiencies, do you think the primary cause of cell death is oxidative stress or ATP insufficiency? What is killing them?

Larsson: I think the bioenergetic deficiency, or perhaps changes in redox state and accompanying secondary effects on metabolism are important.

Spiegelman: Going back to oxidative stress and ROS, I don't recall all the figures in your paper, but since the mutations are going to be random, when you look for oxidative stress as a mass property, I am a bit concerned about that. Since the mutations will be stochastic, and since the probability of generating ROS will depend a bit on where the mutation sits in the electron transport system, did you look at broad sections by immunohistochemistry, looking for individual cells where the hit may have been in a particularly bad place? I can imagine that if you average over a tissue, you could miss something.

Larsson: In principle it could be that ROS production is focal and these are the cells that preferentially are lost. Also, we must remember that this is not the only piece of evidence. The heterozygous SOD2 knockout mouse has more oxidative stress but a normal lifespan.

Spiegelman: It wouldn't be so difficult to go back and look with immunohistochemistry, exactly as you did in muscle. I would predict that depending on where the hits are you might find the occasional cell. By the way, GPX did look like it was up in your cells.

Larsson: GPX is up a little bit. I think what you suggest is quite possible.

Jacobs: You have to remember that every cell contains a large number of different mutant mitochondrial DNA molecules, except to the degree to which in an individual cell or cell type mitotic segregation will during the lifetime purify the population. One thing that is buried in all this work is what is happening in stem cells. Is the ageing phenotype that Nils-Göran Larsson is seeing the result of the gradual loss of viability in stem cells? Not that they are undergoing apoptosis, necessarily, but within certain populations of stem cells particular differentiation programmes may require mitochondrial OXPHOS-derived energy. When a cell then needs to replace another cell by differentiating, it just can't do it.

Larsson: In a way, our results are expected, because a general decrease of respiratory chain function is expected to result in less ROS production.

Spiegelman: Are you addressing the ROS theory of ageing, or are you interested in the effects of the PolG mutation on ROS stress?

Larsson: We have now manipulated mitochondrial function in various tissues of the mouse. We have used different methods. We find very little evidence of oxidative stress. These are the data. It could mean that we haven't done the right experiments or we haven't measured relevant parameters in an appropriate way. However, I think the often supposed direct link between mitochondrial dysfunction and ROS production is not supported by our experiments.

Nicholls: We are very interested to see if we are dealing with oxidative stress or ATP insufficiency in models of glutamate excitotoxicity. We come up with three different models, all of which point towards ATP insufficiency as being the primary cause of neuronal death, rather than oxidative stress. There is the classic study by Heneberry (Novelli et al 1988) which shows the great potentiation of glutamate excitotoxicity by bioenergetic insufficiency.

Brand: If you take the view that it is ATP insufficiency, then this is diametrically opposed to the experiment that Bruce Spiegelman told us about. This is that if you treat with DNP, which presumably causes ATP insufficiency, the system has a robust response, up-regulates the electron transport chain through the transcription factor, and nothing is seen. Why isn't this cutting in and saving these mice?

Larsson: I think the problem is that mitochondrial DNA is messed up.

Lemasters: I was impressed that in the polymerase heterozygotes there was no phenotype. There was no increase in deletions or point mutations. This suggests to me that the mutated mitochondrial DNAs are recognized and removed. A potential mechanism for this would be mitochondrial autophagy. This has a couple of implications. First, not all mitochondrial DNA mutations are alike. mtDNA mutations that make a mitochondrion resistant to autophagy would accumulate. This might occur in a cell-specific manner. Another implication is that individual cells may contain a stem population of mitochondria from which new mitochondria and new mitochondrial DNA are made. A lasting deficit occurs when mitochondrial DNA molecules in stem mitochondria become mutated.

Nicholls: Do we have any evidence for stem mitochondria?

Larsson: There is a much simpler biochemical explanation for this. Replication of mitochondrial DNA is quite slow, and the polymerase has the capacity to correct errors in trans. Let me exemplify by referring to the common PCR reaction. If you do PCR with Taq polymerase, you get a certain level of point mutations. If you add Pfu, which has better proof-reading capacity than Taq, it will correct synthesis errors in trans. Many polymerases have the capacity to go on and off templates. This is the most likely explanation for why we don't see any effects in the heterozygotes. Another explanation could be that there is a difference but our methods aren't sensitive enough to pick it up.

Martinou: There is impaired oxidative phosphorylation in many places in the brain of these mice. If you do not stress the mice, these neurons can survive for months with non-functional mitochondria. For me this is fascinating. These neurons survive by means of glycolysis.

Larsson: At least their capacity to synthesize ATP is much reduced: this is what we find in the knockouts. The intracellular ATP levels are probably normal. They must have secondary down-regulation of processes requiring a lot of ATP. Also, if we force respiratory chain deficient neurons to fire a lot of action potentials by injecting animals with kainic acid there will be massively increased neuronal death.

Nicholls: The simplistic idea we have is that the critical issue is to have enough spare respiratory capacity in mitochondria to deal with a bioenergetic crisis, the '100 year flood'. What Nils-Göran's mice are doing with a nice continuum of increased mitochondrial dysfunction is that they are steadily lowering the 'levees' until they get to the stage that when some form of epileptiform stimulation comes along they don't cope, and die.

Jacobs: It is more dramatic than that. For example, in your MILON mice there is no COX. There is no barrage there at all. This means that they are able to survive on non-respiratory pathways to make ATP. For the demands of a neuron which isn't being subjected to stress, this is sufficient. Have you looked at lactate production in the brains of these mice?

Larsson: No, we haven't looked at this.

Jacobs: This could be another threshold effect: they can clear a certain amount of lactate and survive.

Larsson: The knockout is specific to pyramidal neurons. The glia and other cell types won't be affected by the knockout. There are suggestions that glial cells may feed neurons.

Jacobs: They have to feed them something that they can metabolize.

O'Rourke: Along the lines of substrate supply, it is striking that you maintain complex II, because you are not affecting the nuclear encoded subunits, but you have defects in the other carriers. Are these neurons more sensitive to 3-mercaptopropionic acid (3-MPA) inhibition? This relates to the substantia nigra being particularly sensitive to 3-MPA.

Larsson: We have not done this.

Nicholls: Retaining complex II is completely useless in the whole cell context, because the complex II substrate succinate comes from complex I-dependent dehydrogenases.

O'Rourke: Do they somehow find a way to utilize complex II-derived metabolites?

Nicholls: There is no way this can be done in a cell.

Shirihai: You mentioned that the mice are developing anaemia. This could be a good stem cell model because the erythrocyte differentiation is demanding on mitochondria. It would be interesting to hear how this is influencing anaemia at different stages. One interesting stage is the switch of haemoglobin. This is in the later stage of fetal development.

Larsson: We haven't looked into many of these phenotypes at all. In the knockouts the haematopoiesis isn't down-regulated overall. There are normal levels of the different white blood cell types.

Turnbull: I want to make a comment in relation to the deficient cells. We see these in human brains, both in ageing and in patients with mitochondrial DNA diseases. They are deficient in cytochrome oxidase. Sometimes they accumulate to quite high levels in human brains, so exactly the same is happening in humans as happens in mice. It is quite interesting that Nils-Göran has the ageing phenotype with a random mutagenesis in which there are lots of mutations in a genome, yet the ageing phenotype seen in humans is that associated with clonal expansion. It is a totally different process. It is almost as if the clonally expanded mutations are not the ones that are causing ageing.

Jacobs: We don't know that there isn't clonal expansion.

Larsson: In the heart we find that a subset of the cardiomyocytes are severely respiratory chain deficient. It would be a big step forward if we could generate mice with expansion of clonal mtDNA deletions.

Jacobs: The other phenotype that comes up as the first noticeable phenotype in the mice is male infertility. This is another differentiation pathway that is highly dependent on mitochondrial OXPHOS function. Even though it doesn't quite fit the ageing paradigm, it suggests this explanation.

Spiegelman: The tail of the sperm will need this. It would be interesting if sperm were present but immotile.

Larsson: It is actually quite a dramatic defect in the differentiation process.

Halestrap: We seem to have concluded that it is probably ATP that is important rather than ROS for the cell death pathway. But are we saying that ROS is important in the DNA mutations that upset the respiratory chain, or are they not important at all?

Nicholls: In the simplistic models we work with, we are trying to answer the same question in terms of models of glutamate excitotoxicity. The original idea was that Ca^{2+} floods into post-synaptic mitochondria through NMDA receptors, the mitochondria load with Ca^{2+} and produce ROS, which kill everything. This doesn't work in our model because the ROS is the consequence of the cell death rather than the cause of it. But if you oxidatively damage your mitochondria by some oxidative stress and then give them glutamate, because the oxidative stress has damaged their respiratory capacity, then those cells are more sensitive to glutamate excitotoxicity. An upstream induction of oxidative stress will sensitize the mitochondria to cell death, via a decrease in ATP synthesis.

Halestrap: I follow that; I also want to know what the role of ROS is in DNA mutations.

Larsson: When we sequence mtDNA in these proofreading-deficient mice, we have 65% transitions and 35% transpositions. I spoke to Doug Turnbull last week and he had pretty much the same spectrum when he sequenced human mtDNA. The figures are consistent, indicating that these mutations could be generated by poor proofreading.

Halestrap: What does ROS damage in the DNA do to the proofreading?

Larsson: There is also an issue with 8-hydroxyguanosine. There is a mouse knockout of OGG1, which is a glycosylase that repairs 8-hydroxyguanosine. If the gene is knocked out there is more 8-hydroxyguanosine in the mitochondrial DNA, but no more mtDNA mutations. I am not sure what this means.

Jacobs: One small problem with your point is that in human cells carrying a mutated polymerase with no exonuclease activity in PolG, the pattern of mutations accumulated is different. We have no explanation for this. They are base-directional transitions. In this case the pattern of mutations is not the same as is seen in the mouse. The fact that Doug Turnbull sees a similar pattern of mutations doesn't support the idea that they are replication errors.

Larsson: I think it does. We have done a clean genetic experiment in differentiated tissues of the mouse.

Jacobs: In mouse cells in culture we see the same pattern of mutations as in the whole mouse.

Schon: We talk about ROS as being bad. Peroxide happens to be a signalling molecule and ROS might be good in this context. Patients with mitochondrial pathogenic mutations who presumably have elevated ROS apparently don't accumulate other mutations apart from the one that they have.

Turnbull: That's true, but the evidence that patients with mitochondrial disease have any change in ROS is about as strong as Nils' evidence for the mice.

Larsson: There are exceptions such as Friedreich's ataxia.

Giulivi: Is the pattern of mutations you see with the PolG mutations the same over time?

Larsson: The spectrum is the same but the levels increase with time.

Reference

Novelli A, Reilly JA, Lysko PG, Henneberry RC 1988 Glutamate becomes neurotoxic via the N-methyl-D-aspartate receptor when intracellular energy levels are reduced. Brain Res 451:205–212

Mitochondrial mutations: genotype to phenotype

Eric A. Schon* and Salvatore DiMauro

Departments of Neurology and *Genetics and Development, Columbia University Medical School, 630 West 168th Street, New York, NY 10032, USA

> *Abstract.* Diseases associated with defects of the mitochondrial respiratory chain fall into four major categories: (1) those due to mutations in respiratory chain subunits; (2) those due to mutations that affect respiratory chain assembly; (3) those due to mutations that affect respiratory chain function indirectly, either via alterations in the translation of mtDNA-encoded polypeptides or via alterations in mtDNA integrity; and (4) those due to mutations in nDNA that affect organellar morphology and mobility, in which defects in respiratory chain function can be considered to be 'collateral damage'. All four categories will be discussed.
>
> *2007 Mitochondrial biology: new perspectives. Wiley, Chichester (Novartis Foundation Symposium 287) p 214–233*

Essentially, all human cells contain two genomes: the nuclear genome, consisting of approximately 3 billion bp of DNA (nDNA), encoding approximately 20 000 genes (but specifying a vastly greater number of polypeptides—perhaps 100 000—due to the processes of alternative splicing and alternative translation initiation), and the mitochondrial genome (mtDNA), consisting of 16.6 kb of DNA (mtDNA), encoding only 37 genes, of which only 13 specify polypeptides (the remainder are 2 rRNAs and 22 tRNAs, required for mitochondrial translation) (Fig. 1). Notably, of the ~20 000 nDNA-encoded genes, about 1300 specify polypeptides that are targeted to the mitochondrion's four compartments—the outer membrane (MOM), the inner membrane (MIM), the intermembrane space (IMS), or the matrix. About half of these proteins are required for the organelle's maintenance (e.g. protein import and sorting; protein translation and stability; nucleic acid metabolism; transport of metabolites; response to stress; and the maintenance of organellar morphology), while the other half are required for the organelle's specialized functions (e.g. intermediate, amino acid, and lipid metabolism; respiratory chain and oxidative phosphorylation; signal transduction; and apoptosis) (Table 1).

MUTATIONS

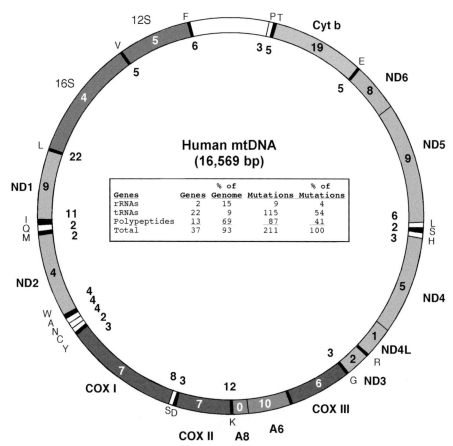

FIG. 1. The human mitochondrial genome. Shown are the 2 rRNAs (12S and 16S), the 22 tRNAs (one-letter amino acid nomenclature), and the 13 polypeptide subunit genes (ND1–6, complex I; Cyt b, cytochrome b subunit of complex III; COX I–III, complex IV; A6 and A8, complex V). Numbers indicate the number of pathogenic mutations found in each gene. The box shows the number and types of the various mutations, compared to the fraction of the genome used by each class of gene.

Our focus here will be on mutations in genes associated with the proper functioning—either direct or indirect—of the respiratory chain-oxidative phosphorylation system (Fig. 2), and a discussion of the relationship between mutated genotypes and pathological phenotypes resulting in respiratory chain disorders. All 13 mtDNA-encoded polypeptides are subunits of this system (Fig. 2), but at least 150 other polypeptides, all nDNA-encoded, are required to build and maintain it in proper working order (Table 2), and all 150 are potential candidates as culprits in respiratory chain disorders of currently unknown aetiology. To make

TABLE 1 Gene products present in human mitochondria

'Maintenance' functions (601)	
Protein translation & stability	223
Carriers & Transporters	118
Nucleic acid metabolism	79
Organellar morphology & inheritance	85
Protein import & sorting	56
Stress response	40
Specialized functions (741)	
Respiratory chain and oxidative phosphorylation	166
Lipid metabolism	120
Signal transduction	112
Apoptosis	97
Intermediate metabolism	95
Amino acid & nitrogen metabolism	61
Miscellaneous/Unknown	90

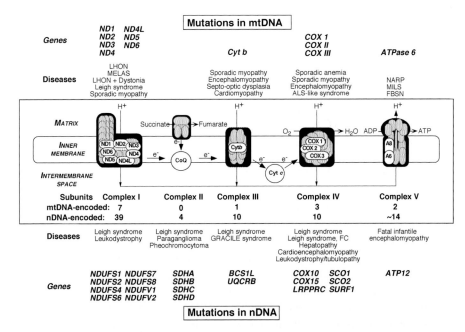

FIG. 2. The respiratory chain. Shown are nDNA-encoded (shaded) and mtDNA-encoded (unshaded) subunits. As electrons (e^-) flow 'horizontally' through the electron transport chain, protons (H^+) are pumped 'vertically' from the matrix to the intermembrane space through complexes I, III, and IV, and then back into the matrix through complex V, to produce ATP. Coenzyme Q (CoQ) and cytochrome c (Cyt *c*) are electron transfer carriers. Relevant genes (in italics) responsible for the indicated respiratory chain disorders are also shown.

TABLE 2 Subunits of the respiratory chain/oxidative phosphorylation system*

	Structural		'Assembly'	
	mtDNA	nDNA	nDNA	Total
Complex I	7	39	3	49
Complex II	0	4	0	4
Complex III	1	10	3	14
Complex IV	3	19	16	38
Complex V	2	17	4	23
CoQ	0	14	1	15
Cytochrome c	0	2	4	6
Heme and FeS metabolism	0	0	17	17
Total	13	105	48	166

*Includes tissue-specific isoforms

matters more complicated, genetic errors causing problems in mitochondrial morphology, fusion, fission and distribution may also result in respiratory chain dysfunction, in a kind of murder at 'six degrees of separation.' For this reason, respiratory chain disorders can be classified in many ways—for example, comparing mutations in mtDNA vs nDNA, or mutations affecting the respiratory chain directly vs. indirectly. Our plan here will be to classify these disorders, which by general agreement have come to be called the 'mitochondrial encephalomyopathies', in four major categories. We will not discuss defects in other mitochondrial systems, such as the TCA cycle or β-oxidation, which are more in the nature of metabolic diseases.

Oxidative phosphorylation (OxPhos) takes the hydrogen present in food (these are the 'reducing equivalents', ultimately protons [H^+], that are associated with the NADH and $FADH_2$ formed in the TCA cycle) to perform one fundamental task: to generate energy as ATP from ADP and inorganic phosphate (ADP + Pi → ATP). The production of ATP is not due to any direct biochemical reaction with H^+, but rather is driven by a proton gradient across the inner mitochondrial membrane (IMM) through ATP synthase, a lollipop-shaped molecular motor that uses the chemical energy of the H^+ gradient to drive a rotating camshaft (the 'stick') that, in turn, drives three conformations of the business end of the machine (the 'candy') to allow for ADP+Pi binding, ADP+Pi reaction, and ATP release. As the generation of protons (H^+) from neutral hydrogen requires the generation of an equal number of electrons (e^-), two necessary side reactions are the oxidation of organic compounds to produce CO_2 in the various steps of intermediate metabolism, and the reduction of oxygen to water ($H_2 + 2O_2 \rightarrow 2H_2O$) via the respiratory chain (Fig. 2). In the electron-transport chain, NADH and $FADH_2$ transfer electrons to ubiquinone (also called coenzyme Q10 [CoQ10]) via complexes I and II, respectively, and from there to complexes III and IV via a second electron carrier, cytochrome c. It is at the terminus of the respiratory chain at complex IV (also called cytochrome c oxidase, or COX) where the electrons are finally reunited with those protons not destined to drive ATP synthesis to form water (besides moving electrons, complexes I, III, and IV also pump the protons that will be used to make ATP from the matrix to the IMS). Eventually, we exhale the H_2O and CO_2.

Mitochondria obey distinctive rules of population genetics. First, they are maternally inherited. Thus, most pathogenic mtDNA mutations are transmitted from a woman to all her children, but only her daughters will pass it on to their children. Second, mtDNAs are present in hundreds or thousands of copies per cell (polyplasmy) and mutations are often *heteroplasmic*, that is, normal and mutated mtDNAs coexist in the same cell or tissue. Third, a minimum number of mutant mtDNAs is required to impair function and cause symptoms (the *threshold effect*), with the threshold depending on the energetic demands of the tissue in question. Fourth, the proportion of mutant mtDNAs can vary both in space and time (because mtDNAs are parsed stochastically during cell division); such *mitotic segregation* can cause the clinical presentation to change over time (DiMauro & Schon 2003).

Mitochondrial respiratory chain diseases

Because the construction of the respiratory chain and oxidative phosphorylation system is a collaboration between the nuclear and mitochondrial genomes, patients with diseases affecting this system show a bewildering array of signs and symptoms that can be maternally inherited (mutations in mtDNA), Mendelian-inherited (mutations in nDNA), or even sporadic (spontaneous mutations in either genome). This problem is compounded by the population nature of mitochondrial genetics and the stochastic nature of mtDNA replication and organellar inheritance. Taken together, the problem of classification of mitochondrial respiratory chain diseases has been a daunting one. We will take a four-pronged approach here. Due to space limitations, the classification is not intended to be comprehensive (there are a number of excellent detailed reviews on the topic; DiMauro & Hirano 2005), but rather to be illustrative of principles and trends that have emerged in the last few years.

Mutations in respiratory chain subunits

As noted above, the respiratory chain-oxidative phosphorylation system consists of five multi-subunit complexes (almost 120 polypeptides [including tissue-specific isoforms of individual subunits], of which 13 are mtDNA-encoded) and two electron carriers, CoQ and cytochrome c (Table 2).

Mutations have been identified in 12 of the 13 mtDNA-encoded polypeptides (only ATPase 8 has been spared) (see Fig. 1). Because most of the mutations are heteroplasmic, the clinical presentations are highly variable. Some mutations cause pure myopathies (e.g. many mutations in cytochrome b, which, for unknown reasons, often arise sporadically, with no evidence of maternal inheritance). Others, invariably mutations in ND genes, cause Leber hereditary optic neuropathy (LHON), an adult-onset form of blindness that, mysteriously, affects men more often and more severely than women. Others cause encephalomyopathies, cardiomyopathies, or are multisystem disorders whose severity depends on the mutation load.

At the severe end of the spectrum is Leigh syndrome (LS), a fatal encephalopathy of infancy: to date LS due to errors in mtDNA has been associated only with mutations in ND2, ND4, and ND6 of complex I, and in subunit A6 of complex V. In the latter case, the mutation, when present in high amounts (>90%) causes maternally-inherited Leigh syndrome (MILS), but at lower mutant loads (70–90%) a different disorder arises in later life, called neuropathy, ataxia, and retinitis pigmentosa (NARP). The difference between NARP and MILS illustrates a key hallmark of mtDNA diseases, namely, that different mutant loads can result in completely different clinical presentation.

Interestingly, mutations in nDNA-encoded subunits of the respiratory chain have been found only for subunits of complexes I and II. All the mutations are recessive and typically cause LS. Of the four subunits of complex II, only mutations in the flavoprotein subunit (SDHA) cause LS (Bourgeron et al 1995). Mutations in the other three subunits (SDHB, C, and D) have been found, but surprisingly, they all result in tumours—paragangliomas and pheochromocytomas (Dahia 2006). Perhaps the failure to find mutations in complexes III–V is due to complexes I and II running 'in parallel' to provide electrons to CoQ, so that complex II can compensate for loss of complex I, and vice versa. On the other hand, complexes III–V run 'in series' downstream of CoQ, so that a mutation that impairs electron flow in these complexes will likely be fatal (presumably, detection of patients with mutations in mtDNA-encoded structural subunits is due to heteroplasmy; homoplasmy would result in embryonic lethality). Of course, given the number of subunits in these complexes, an exception to this 'rule' will almost certainly be found one day.

No mutations have been found in cytochrome c, perhaps because loss of cytochrome c would be incompatible with life, but mutations in two CoQ10 biosynthetic enzymes have recently been identified, causing infantile encephalomyopathy and nephrotic syndrome (Quinzii et al 2007).

Mutations that affect respiratory chain assembly

A growing number of disorders are due to mutations in 'ancillary' proteins that are not subunits of any complex. Rather, they are required for the proper assembly of the respiratory complexes, including the incorporation of prosthetic groups. Most of the assembly defects were identified in patients with isolated COX deficiency, many with LS, in whom all 13 COX subunit genes appeared to be normal. Many of the COX assembly genes in yeast (most of which were identified by Alex Tzagoloff at Columbia) have homologues in humans, and a number of these have now been found to cause COX deficiency in infancy (Pecina et al 2004). The first such mutation, causing COX-deficent LS, was found in 1998 in a gene called *SURF1*. A second form of COX-deficient LS, prevalent in the Saguenay-Lac Saint-Jean region of Quebec (LS-French-Canadian type, or LSFC), is due to mutations in *LRPPRC*, the human homologue of *PET309*, a gene required for maturation of COX1 mRNA in yeast (Xu et al 2004). Mutations in COX10 (a protohaem: haem-*O*-farnesyl transferase, involved in the biosynthesis of haem *a*) have been found in patients with COX deficiency presenting with leukodystrophy and tubulopathy, deafness and cardiomyopathy, and LS. Mutations in COX15, another enzyme required for haem *a* biosynthesis also cause LS. Two other complex IV assembly factors are related proteins: SCO1 and SCO2. Both appear to be required for the insertion of copper into the COX holoprotein, but there is some evidence, based on structural grounds, that the SCO proteins may also be involved in mitochondrial redox sensing (Williams et al 2005). Mutations in SCO1 cause a fatal hepato-encephalomyopathy and those in SCO2 a fatal cardio-encephalomyopathy, both with isolated COX deficiency. The reason for the tissue specificity of the two disorders is unknown.

Mutations have now been found in assembly genes for other complexes. Mutations in a complex I assembly gene *(B17.2L)* caused severe cavitating leukoencephalopathy (Ogilvie et al 2005). Mutations have been found in two complex III assembly genes. Errors in BCS1L, required for assembly of the Rieske FeS protein, causes GRACILE syndrome (growth retardation, aminoaciduria, cholestasis, iron overload, lactic acidosis and early death) (Visapaa et al 2002); mutations in UQCRB (ubiquinone-binding protein; QP-C subunit or subunit VII) caused complex III deficiency with hypoglycaemia and lactic acidosis (Haut et al 2003). Finally, a mutation in ATPAF2 (also called ATP12), a compex V assembly protein that binds the a subunit, caused a fatal encephalomyopathy in two infants (De Meirleir et al 2004).

Mutations that affect respiratory chain function indirectly

This group of mutations actually falls into two subcategories: those that are due to alterations in the integrity or plasticity of mtDNA and those that result in

MUTATIONS 221

alterations in the translation of mtDNA-encoded polypeptides. In many patients with mutations in this category, the muscle biopsy shows segmental regions of massive proliferation of mitochondria, called ragged-red fibres (RRF) (when stained with modified Gomori trichrome) or 'ragged-purple' fibres (when stained for succinate dehydrogenase [SDH] activity). Because the RRF contain a high proportion of mutated mtDNAs, they usually lack cytochrome c oxidase (COX) activity.

Besides 13 polypeptides, the mitochondrial genome encodes 2 rRNA and 22 tRNAs, and mutations in these latter 24 genes also cause disease. One would think that because all 24 genes are required for translation of the mt-mRNAs, mutations in any of them would cause pretty much the same disorder (at an equivalent mutation load, of course). In fact, the opposite is often true: mutations in specific tRNAs and rRNAs cause remarkably different, and often stereotypical disorders. Four genes illustrate this phenomenon best. (1) Mutations in $tRNA^{Leu(UUR)}$ are associated with MELAS (mitochondrial encephalomyopathy, lactic acidosis, and stroke-like episodes). (2) Mutations in $tRNA^{Lys}$ are associated with MERRF (myoclonus epilepsy with RRF). (3) Mutations in $tRNA^{Ile}$ are associated with isolated cardiomyopathies. (4) Mutations in 12S rRNA are associated with aminoglycoside-induced non-syndromic deafness (and sometimes with cardiopathy). While one can speculate about why mutations in different genes cause different syndromes, the fact remains that mutation- and tissue-specificity is *terra incognita*—we really have no understanding of this phenomenon, although some inroads into the problem are beginning to be made (Antonicka et al 2006).

Large-scale deletions of mtDNA (Δ-mtDNA) stand in a category of their own. The deletions range in size from approximately 2–10 kb, and arise spontaneously (i.e. there is no evidence of maternal inheritance). These spontaneous deletion disorders typically cause one of three syndromes: (1) a pure myopathy called progressive external ophthalmoplegia (PEO); (2) a fatal multisystem disorder called Kearns-Sayre syndrome (KSS); and (3) a haematopoietic disorder called Pearson syndrome (PS). As all three are sporadic, and as different patients have different 'unique' deletions, it is presumed that the mtDNA deletion in each patient arose in a single 'primordial' molecule, either in the mother's germline (thereby affecting all tissues and causing KSS) or early in embryogenesis (thereby affecting, for example, only mesodermal lineages and causing PEO). While these giant deletions obviously remove structural genes encoding respiratory chain subunits, *a fortiori* they also remove tRNAs. It appears to be the loss of tRNAs—meaning the inability to translate all 13 mt-mRNAs (whether located inside the deleted region or not)—that determine the clinical outcome. This is why different KSS patients can have different deletions and yet present with a similar phenotype (Nakase et al 1990).

A major cause of mitochondrial disease is due to Mendelian-inherited errors in genes that impair mtDNA integrity. These mutations result in either the generation of *multiple* large-scale deletions during the lifetime of the patient, or the near-total absence of mitochondrial genomes (mtDNA depletion) in some tissues of affected patients, or both. Patients with multiple mtDNA deletion syndromes have ocular and limb myopathy (PEO, ptosis, proximal weakness), and most of them also have involvement of other systems, including brain (ataxia, dementia, psychoses), eye (cataracts), ear (sensorineural hearing loss) and peripheral nerves (sensorimotor neuropathy).

Notably, nearly all causes of both multiple deletions and depletion are due to mutations in genes needed to replicate mtDNA, including polymerases, helicases, and genes required for the maintenance of mitochondrial nucleotide pools (Van Goethem 2006). These include the adenosine nucleoside translocator (*ANT1*), the two mitochondrial polymerase γ subunits—the catalytic (*POLG*) and regulatory/processivity (*POLG2*) subunits—and a helicase called Twinkle (*PEO1*). In addition, nucleotide pools are affected by mutations in mitochondrial thymidine kinase (*TK2*) and deoxyguanosine kinase (*DGUOK*), in the β subunit of succinyl coenzyme A synthetase (*SUCLA2*), and in the cytosolic enzyme thymidine phosphorylase (*TP*).

Mutations in *TP* are of particular interest, as they cause a devastating autosomal recessive multisystem disease of young adults called MNGIE (mitochondrial neurogastrointestinal encephalomyopathy) (Hirano et al 2004). MNGIE is characterized clinically by PEO, leukoencephalopathy, neuropathy, and most notably, intestinal dysmotility leading to cachexia and early death. Loss of TP activity causes multiple deletions, depletion, and even mtDNA point mutations in muscle, even though TP is not expressed in this tissue. TP normally converts thymidine to thymine, but in patients deoxyuridine and thymidine, both of which are toxic, accumulate massively in the blood. Reduction in thymidine via bone marrow transplantation may be a viable therapeutic possibility (Hirano et al 2006).

Mutations in *POLG* can be inherited as both autosomal-dominant and recessive traits. Patients (mainly adults) usually have PEO (they have multiple mtDNA deletions in muscle), but they also present with a wide spectrum of symptoms, including peripheral neuropathy, ataxia, psychiatric disorders, parkinsonism, myoclonus epilepsy and gastrointestinal symptoms mimicking MNGIE. Some recessive mutations in *POLG* cause Alpers syndrome, a severe hepatocerebral disorder in children associated with mtDNA depletion and extreme vulnerability to valproate administration. For unknown reasons, most Alpers children are compound heterozygotes, with at least one mutation in the 'linker' region connecting the polymerase and exonuclease (i.e. proofreading) domains, whereas the less severely-affected adults with PEO usually have mutations only in the polymerase domain.

Beside POLG, hepatocerebral syndrome is also caused by mutations in two other genes: deoxyguanosine kinase (*DGUOK*), and a new gene of unknown function called *MPV17* (Spinazzola et al 2006). Surprisingly, the same homozygote mutation in *MPV17* was found in a southern Italian family and in the Navajo population of the southwestern USA (where it causes Navajo neurohepatopathy, or NNH (Karadimas et al 2006). While the function of MPV17 is unknown, it may be significant that in immunohistochemistry, the pattern of MPV17 staining appears to be more punctate than 'spaghetti-shaped' (Spinazzola et al 2006), similar to the pattern seen with other mtDNA-binding proteins that are associated with nucleoids (aggregates of mtDNA and associated proteins), such as POLG and PEO1. This result implies that MPV17 may in fact be required for regulating some aspect of mtDNA plasticity.

Proper and correct translation of the 13 mtDNA-encoded subunits of the respiratory chain requires several factors for ribosomal initiation, elongation, and termination, all of which are encoded by nDNA. As might be expected, defects arising from errors in mitochondrial translation of mt-mRNAs cause severe combined defects of all four respiratory chain complexes containing mtDNA-encoded subunits (i.e. complexes I, III, IV, and V) (Jacobs & Turnbull 2005). From the biochemical point of view, errors in translation look essentially identical to mtDNA depletion. Thus, translation errors should be suspected in infants or children with hepatocerebral syndrome, encephalopathy, or cardiomyopathy and multiple respiratory chain defects but who have no evidence of mtDNA depletion.

To date, mutations in four nuclear genes have been described. Two of them affect two of the four known mitochondrial elongation factors, G1 (*GFM1*; also called EFG1), and Ef-Ts (*TSFM*) (the other two are Ef-G2 [*GFM2*] and Ef-Tu [*TUFM*]). The third mutation affects *MRPS16*, encoding one of the small ribosomal protein subunits. The last mutation does not affect ribosomal translation *per se*, as it is in a tRNA-modification gene, called *PUS1*, which is required for the pseudouridylation of mitochondrial tRNA genes. This disorder has been called MLASA, because it is characterized by myopathy, lactic acidosis, and sideroblastic anaemia.

Mutations in nDNA that affect organellar morphology and mobility

Defects in mitochondrial fusion, fission, and movement are a relatively new area of interest both from an academic and clinical point of view. Because the network of hundreds or thousands of mitochondria comprise a decentralized or distributed power system, organelles need to move throughout the cell to deliver ATP when and where it is most needed. In order to move, mitochondria bind to, and travel along, microtubules, both in the anterograde (from the cell body, using kinesins) and retrograde (towards the cell body, using dyneins) directions. The importance of efficient organellar movement is most obvious in neurons, the longest cells in the body, where mitochondria need to travel vast distances to deliver the ATP required, for example, for synaptic transmission. Defects in respiratory chain function due to alterations in mitochondrial topology and distribution can be considered to be 'collateral damage' (Chan 2006).

The first defect in mitochondrial motility was found in a family with autosomal dominant hereditary spastic paraplegia type 10 (SPG10) harbouring mutations in kinesin heavy chain 5A (KIF5A). Interestingly, the mutation affects a region of the protein involved in microtubule binding. Notably, KIF5A interacts with two kinesin-associated mitochondrial adaptor proteins that connect mitochondria to kinesins, called TRAK1 (trafficking kinesin-binding protein 1, also called OIP106) and TRAK2 (also called GRIF1). An autosomal-recessive form of HSP type 20 (SPG20, also called Troyer syndrome) is caused by mutations in spartin, a microtubule-binding protein that also binds mitochondria. Finally, autosomal-recessive HSP type 7 (SPG7) is due to mutations in paraplegin, a MIM-localized metalloprotease. Mice deficient in SPG7 had axonal degeneration and defective axonal transport associated with dysfunctional mitochondria. Remarkably, the yeast homologue of SPG7, Yta12, is required to process the ribosomal protein Mrpl32 for assembly into the mature mitoribosome (Nolden et al 2005), potentially bringing us full circle by connecting mitochondrial mobility defects to a more global problem in mitochondrial translation.

At least three proteins are required for mitochondrial fission: dynamin-related protein (DNM1L), fission-related protein (FIS1), and mitochondrial division protein (Mdv1 in yeast; probably FBXW7 in human), and two for mitochondrial fusion: mitofusins 1 (MFN1) and 2 (MFN2), both homologues of the *Drosophila* 'fuzzy onion' protein. In addition, OPA1, a MIM protein, helps establish 'contact sites' between the MOM and IMM in order to co-ordinate organellar fusion and fission. It turns out that mutations in OPA1 cause autosomal dominant optic atrophy (the Mendelian counterpart of LHON, as it were), while mutations in MFN2 cause an autosomal dominant axonal variant of Charcot-Marie-Tooth disease type 2A (CMT2A). Finally, mutations in ganglioside-induced differentiation protein 1 (GDAP1), a MOM protein that interacts with mitofusins to regulate the mitochondrial network, cause CMT4A, an autosomal recessive, severe, early-onset form of either demyelinating or axonal neuropathy (Pedrola et al 2005).

Acknowledgements

This work has been supported by grants from the National Institutes of Health (NS11766 and HD32062), from the Muscular Dystrophy Association, from the Marriott Mitochondrial Disorder Clinical Research Fund (MMDCRF), and from Edison Pharmaceuticals.

References

Antonicka H, Sasarman F, Kennaway NG, Shoubridge EA 2006 The molecular basis for tissue specificity of the oxidative phosphorylation deficiencies in patients with mutations in the mitochondrial translation factor EFG1. Hum Mol Genet 15:1835–1846

Bourgeron T, Rustin P, Chretien D et al 1995 Mutation of a nuclear succinate dehydrogenase gene results in mitochondrial respiratory chain deficiency. Nat Genet 11:144–149

Chan DC 2006 Mitochondria: dynamic organelles in disease, aging, and development. Cell 125:1241–1252

Dahia PL 2006 Evolving concepts in pheochromocytoma and paraganglioma. Curr Opin Oncol 18:1–8

De Meirleir L, Seneca S, Lissens W et al 2004 Respiratory chain complex V deficiency due to a mutation in the assembly gene ATP12. J Med Genet 41:120–124

DiMauro S, Schon EA 2003 Mitochondrial respiratory-chain diseases. N Engl J Med 348:2656–2668

DiMauro S, Hirano M 2005 Mitochondrial encephalomyopathies: an update. Neuromuscl Disord 15:276–286

Haut S, Brivet M, Touati G et al 2003 A deletion in the human QP-C gene causes a complex III deficiency resulting in hypoglycaemia and lactic acidosis. Hum Genet 113:118–122

Hirano M, Nishigaki Y, Marti R 2004 Mitochondrial neurogastrointestinal encephalomyopathy (MNGIE): a disease of two genomes. Neurologist 10:8–17

Hirano M, Marti R, Casali C et al 2006 Allogeneic stem cell transplantation corrects biochemical derangements in MNGIE. Neurology 67:1458–1460

Jacobs HT, Turnbull DM 2005 Nuclear genes and mitochondrial translation: a new class of genetic disease. Trends Genet 21:312–314

Karadimas CL, Vu TH, Holve SA et al 2006 Navajo neurohepatopathy is caused by a mutation in the *MPV17* gene. Am J Hum Genet 79:544–548

Nakase H, Moraes CT, Rizzuto R et al 1990 Transcription and translation of deleted mitochondrial genomes in Kearns-Sayre syndrome: implications for pathogenesis. Am J Hum Genet 46:418–427

Nolden M, Ehses S, Koppen M et al 2005 The m-AAA protease defective in hereditary spastic paraplegia controls ribosome assembly in mitochondria. Cell 123:277–289

Ogilvie I, Kennaway NG, Shoubridge EA 2005 A molecular chaperone for mitochondrial complex I assembly is mutated in a progressive encephalopathy. J Clin Invest 115:2784–2792

Pecina P, Houstkova H, Hansikova H et al 2004 Genetic defects of cytochrome *c* oxidase assembly. Physiol Res 53:S213–S223

Pedrola L, Espert A, Wu X et al 2005 GDAP1, the protein causing Charcot-Marie-Tooth disease type 4A, is expressed in neurons and is associated with mitochondria. Hum Mol Genet 14:1087–1094

Quinzii CM, DiMauro S, Hirano M 2007 Human coenzyme Q10 deficiency. Neurochem Res 32:723–727

Spinazzola A, Viscomi C, Fernandez-Vizarra E et al 2006 MPV17 encodes an inner mitochondrial membrane protein and is mutated in infantile hepatic mitochondrial DNA depletion. Nat Genet 38:570–575

Van Goethem G 2006 Autosomal disorders of mitochondrial DNA maintenance. Acta Neurol Belg 106:66–72

Visapaa I, Fellman V, Vesa J et al 2002 GRACILE syndrome, a lethal metabolic disorder with iron overload, is caused by a point mutation in BCS1L. Am J Hum Genet 71:863–876

Williams JC, Sue C, Banting GS et al 2005 Crystal structure of human SCO1: implications for redox signaling by a mitochondrial cytochrome *c* oxidase "assembly" protein. J Biol Chem 280:15202–15211

Xu F, Morin C, Mitchell G et al 2004 The role of the LRPPRC (leucine-rich pentatricopeptide repeat cassette) gene in cytochrome oxidase assembly: mutation causes lowered levels of COX (cytochrome *c* oxidase) I and COX III mRNA. Biochem J 382:331–336

DISCUSSION

Larsson: I disagree with your interpretation of the experiment in which you take two different mtDNA deletions and introduce them into the same cell to recover a chimeric, recombined, molecule. You say that this is evidence for homologous recombination. I am not convinced. Whenever DNA is replicated the DNA replicating enzyme may jump from one template to another, and generate chimeric molecules. We get this all the time with PCR reactions. Homologous recombination involves specific enzymes whereas copy choice recombination is a process where the polymerase switches from one template to another. Your experiment doesn't give any frequency of the recombination event. It could be an extremely rare event that is amplified in cloned cells. Personally, I will not be convinced that homologous recombination exists in mammalian mitochondria until someone identifies the mitochondrial enzymes that must be present if there is homologous recombination in mammalian mitochondria. There are many reports identifying recombinant mtDNA molecules by PCR. I find these reports rather unconvincing.

Schon: That was a Southern blot.

Larsson: Still, this could be an extremely rare event that has been amplified. Do you have any idea of the frequency?

Schon: I agree that the Holliday junction resolvase has never been found in mammalian mitochondria. This is a real problem. Nor has RecA been found.

Larsson: We spent quite a bit of time looking for these enzymes but couldn't find them. It doesn't mean that they don't exist.

Schon: If I had got up here and said that I'd found Holliday junction resolvase, you'd be telling me that you won't believe any functionality until I see a recombed molecule! Perhaps we are dealing with semantics here.

Larsson: It is not semantics but rather a matter of biochemical definition.

Shirihai: Can you put this together with Gresham's law? I thought your comment was that the level of heteroplasmy stays stable in post-mitotic cells.

Schon: That's not Gresham's law. That is not Muller's ratchet, because that cell was put under incredible selective pressure to couple. This is also under pressure to select for function. My comment earlier in the meeting about Gresham's law had to do with population genetics, and random genetic drift that slowly allows one kind of genotype to accumulate and rarely goes back to the other genotype under the absence of selective pressure. Muller's ratchet is talking about genetic drift. In this case we are trying to get the molecules to reduce their heteroplasmy because this will prevent the cells from maintaining themselves.

Spiegelman: In these ragged-red fibres, is this a fibre-type switch? What kinds of muscle fibres are sensitive to this?

Schon: Type 1s. The ones that are sick are type 1s.

MUTATIONS   227

Turnbull: There are cytochrome oxidase deficient type 1s or type 2s. It is not just type 1s. The ragged-reds with mitochondrial proliferation tend to be more type 1s because they tend to have more mitochondria to start with. But the original finding of mosaic cytochrome oxidase deficiency is either type 1 or type 2.

Spiegelman: We have data that in certain kinds of mitochondrial problems, we have used uncoupling agents that have induced PGC1α, which causes a switching to type 1 and 2A fibres. Is the red in ragged-red fibres actually due to firing off of this PGC1α conversation?

Jacobs: It is red because of the histological staining.

Schon: I tried to get someone in our group to do immunostaining for PGC1. It should be done.

Spiegelman: It is hard to do. Because the protein turns over so fast, there is less than you'd imagine from the RNA levels. It is doable, but it's a lot easier with *in situ* hybridisation.

Orrenius: I like your SCO work. SCO2 is known to be regulated by p53. Are both SCO genes regulated by p53?

Schon: No. Only SCO2, which only has one intron. This was shown recently. p53 does not bind to SCO1.

Orrenius: How important is p53 for SCO2 activity and for respiratory activity? As you might know, it has been referred to as a contributor to the Warburg effect. If p53 is non-functional, there would be a down-regulation of respiratory activity and, hence, ATP production by oxidative phosphorylation. Another recently discovered p53 target gene is TIGAR, which seems to halt glycolysis while stimulating NADPH generation via the pentose phosphate shunt. Down-regulation of these two genes in p53-mutant tumour cells might contribute to the Warburg effect (see Orrenius et al 2007 for review).

Schon: That is right. p53 is positively correlated with SCO. So if p53 is down, SCO is down and respiratory chain is down. p53 is inversely correlated with TIGAR, so when it is down TIGAR is up and glycolysis is up. Working together, it seems that this is the beginning of the explanation of the Warburg effect. We are ready to revisit the Warburg. One other aspect I should raise is that we have a project in the lab trying to understand ragged red fibre formation. Stephanie Zanssen has shown that oncocytomas, which are benign tumours that are loaded with mitochondria, have stereotypical reciprocal chromosomal translocations. A fair number of them always translocate on chromosome 11. She has mapped three of them, and they all map upstream of cyclin D1. We have to be careful, because two things are happening in an oncocytoma: the cell is proliferating and the mitochondria are proliferating. But this may be kicking off a cellular and mitochondrial proliferation. Carlos Moraes found a connection between this and caveolin 1. It is a complicated story. Now that we know things about mitochondria that we didn't know in 1956 we might be able to interdict cellular function by the Warburg effect that we could not have done 40 years ago.

Nicholls: You talk about SCO2 as being a potential redox sensor. What exactly would it sense? What sort of mid point potential does it have?

Schon: We haven't done that experiment yet. We don't even know how it would work, because if it senses it must report. It would be hard to imagine that it reported to the nucleus, since it is inside mitochondria, unless it reported through a small molecule or its protruding N-terminus.

Rich: For it to be a redox reporter, it must have a redox group, unless you imagine a really exotic mechanism involving different conformations that are changed by other redox proteins. Could the redox group be the copper itself, or are there any other redox groups in your structure such as disulfides?

Schon: The binding of the copper is through disulfide. It is two cysteines. It could be the disulfide group itself that may make an allosteric modification to the SCO2 structure.

Rich: In your X-ray crystal structure, is there a disulfide or is it two SH groups?

Schon: The latter.

Rich: It is interesting that you lose the copper. It looked from your spectrum that it was binding copper until you add H_2O_2. This is probably just causing oxidative damage to the protein, making it fall apart a bit. In addition, in the X-ray crystal structure redox proteins are prone to reduction by the X-ray beam. Could the beam be reducing your Cu^{2+} to Cu^+? This type of behaviour could be what is wanted for a transport protein i.e. strong binding to one redox state and weak binding to another so that it can bind strongly, then release the metal by change of redox state.

Schon: I am not a crystallographer, so I can't respond. The crystallization is done in PEG. It turned out that it was old PEG so there was peroxide in the crystallization medium. We then had to recrystallize it, fresh, and we still didn't find copper in the crystal.

Lemasters: Will people on a ketogenic Atkins diet get young again?

Schon: No, I don't think so. First of all, this is not the Atkins diet. It is a diet that is given to epileptic children. We are giving a high fat diet to the cells. We have obtained approval to get one or two patients in to see whether we can shift heteroplasmy. You could imagine that this might be bad: if you have a homoplasmic neuron and put a patient on this diet you might kill that neuron. I would like to do this in a patient with PEO, where it is confined to muscle.

Turnbull: The ketogenic diets used in children with intractable epilepsy are quite effective. I agree that you wouldn't want to do it with anyone with CNS involvement, although unless you are stimulating their neurons you are probably OK. I'd like to comment on the MELAS data. It's the commonest mitochondrial DNA mutation in our cohort. We have done a bit of work on the vascular nature of this, and smooth muscle seems to have an incredibly high level of mutations. We find deficient smooth muscle cells present in the brain, muscle and the gut. These people get horrendous bowel problems. I am not sure about your NO hypothesis. Because this is a heteroplasmic mutation which seems to segregate specifically in specific tissues, then it might all be due to the mitochondrial segregation.

Rizzuto: The MELAS paradox intrigued me. What is the explanation for the COX positivity and negativity of the tumour?

Schon: It is a threshold. There is a disease called maternally inherited progressive external ophthalmaplegia caused by the same mutation. In those patients there are COX-negative ragged-red fibres. If we do single fibre PCR, MELAS patients have on average in COX-positive ragged-red fibres between 90% and 95% mutation, and PEO patients with COX-negative ragged-red fibres have between 95% and 98% mutations. It is a small difference, but this is the threshold that needs to be passed to get negativity.

Rizzuto: Regarding the NO sponge idea, there is an example where this can be tested—inflammation. NO is produced and the extent of oedema and redness is proportional to the NO. Following this idea, MELAS patients compared to the MERFF should have less inflammation.

Turnbull: I don't think there's any evidence that these patients have altered inflammatory responses.

Halestrap: We have mentioned the Warburg effect. Is there a correlation between rapid cell division and not using mitochondria to provide ATP? Transformed cells divide rapidly and they are glycolytic. We know that things are happening to mitochondria during cell division and that during the cell cycle, glycolysis is switched on. Is there something that means the mitochondria must be switched off while they are going through processes such as fission that are necessary for cell division?

Schon: We always say that mitochondrial replication and DNA replication are unrelated to the cell cycle. This can't be right. There is one paper that says there is a cell cycle checkpoint that waits for mitochondrial function. If this is not up to scratch, you don't go through the checkpoint (Mandal et al 2005). This can't be right in patients. Perhaps in normal circumstances, this does happen. There is a ciliate that only has one mitochondrion, and in that organism there is checkpoint control.

Halestrap: If you force cells to be oxidative, as you remove glucose and give them something like butyrate, they tend to undergo apoptosis. This happens in gut cells, and is one of the reasons that short-chain fatty acids are good for gut carcinomas (Singh et al 1997). You said that in your ketogenic diet the cells didn't proliferate, which is consistent with this.

Schon: We didn't check the rate of apoptosis. They weren't dividing at all.

Halestrap: Does a cell have to be glycolytic to divide?

Schon: Let's be careful. Humans always make glucose. This is one of the problems we will have when this patient comes in. We have added back glucose with ketones present to see what the threshold is for getting this to happen. It is about 2 mM, which is just at the border of patients who are fasting.

Scorrano: I have a question for Jean-Claude Martinou. If I remember correctly, if you switch the cell medium to galactose, mitochondria are more elongated.

Martinou: If I remember correctly, the cells divide less than when they are cultured in the presence of glucose, but they still divide.

Schon: We couldn't get this to happen in galactose. It was too weak.

Martinou: If we prevent the fission of mitochondria, then the cells stop in G1. There are two explanations. Either fission is necessary as a checkpoint for S phase. Or (less interestingly) preventing fission damages the mitochondria, which produce less ATP. Therefore the cells stop in G1.

Halestrap: There is a new inhibitor that was developed as an immunosuppressant, and it stops T lymphocyte activation. It is incredibly potent, and its target is the lactate transporter (Murray et al 2005). When T cells go into this proliferative stage they get a massive up-regulation of glycolysis and lactate transport. If you stop that they can't proliferate. This is another example of when you want to proliferate fast, you have to be glycolytic.

Schon: When Giovanni Manfredi was in our lab he did a nice experiment. We wanted to ask whether ATP is compartmentalized in the cell. So we took luciferase and targeted it to different compartments, one of which was the nucleus. It turned out that most of the ATP that is in the nucleus is coming from mitochondria, not from glycolysis.

Nicholls: How does it get there without equilibrating with the cytoplasmic pool?

Schon: The way we did this is since we are targeting to a specific compartment, the luciferase light is stoichiometric to the amount of ATP. Then we added inhibitors of glycolysis or mitochondria, and watched the light go up and down. When glycolysis was inhibited the light didn't go out that much.

Nicholls: This is the opposite of the Pasteur effect. If oxidative phosphorylation is switched off, glycolysis goes way up, and the other way round. I am not sure it is that clean.

Shirihai: If at that time cells are glycolytic, would you argue that the oligomycin reversal point test will show that all mitochondria should depolarize if that was correct? This means that if you add oligomycin to the culture, you see a depolarized cell wherever they are proliferating, and we don't.

Halestrap: That is different. It doesn't mean that they can't maintain their membrane potential; it is just that the mitochondria are not the major source of ATP for the cell. They could still be oxidizing substrates but they can't be producing a lot of ATP to drive the proliferation. For some reason the cells have to use glycolysis for their bulk ATP. I think it is probably because you are doing things to mitochondria and you don't want them to get tied up in all this bioenergetic activity.

Nicholls: I have a general question. Why do Orian Shirihai's method for getting dysfunctional mitochondria out of the system and John's mitochondrial autophagic system for getting hold of damaged mitochondria tragically fail in these mitochondrial diseases? Why is heteroplasmy not self-correcting by these removal methods?

Jacobs: An obvious explanation is that the damage is not in a single mitochondrion, but is throughout the mitochondria of the cell. The damaged DNA may be in a specific place with the mutation, but the effect could be global. Therefore just turning over a population of the mitochondria is not going to resolve the issue one way or another.

Nicholls: Is there heteroplasmy within the individual mitochondria?

Jacobs: If there is no fusion or fission, if things are that bad then under those circumstances you could lose a population of mitochondria that don't have a functional genome. But if you have a subfunctional genome distributed throughout the cell and contributing to the proteins in the mitochondrial network in the cell, then just turning over a population of the mitochondria that happened to be where the protein damage is occurring isn't enough. Another issue of course, is whether the protein damage that we are talking about is the same as not having a functional OXPHOS system. Everything that Nils-Göran Larsson presented in his paper suggests that this is not the case.

Youle: Is the cell perhaps trying to do that? Maybe the steady-state levels are the same, but is the turnover rate of proteins higher in the cells of patients?

Schon: Unassembled proteins disappear.

Shirihai: There is an important point which differentiates removal by autophagy from the combination of mitochondrial dynamics + autophagy. Mitochondrial autophagy is targeting those with a very low membrane potential. Autophagy is selecting based on a yet unknown criterion of minimal activity (for example membrane potential or ATP production), say 30% of normal activity. If the only mechanism for quality maintenance was autophagy, we would have ended up with a cell which is set very close to this criterion, i.e. 30% activity. This is where mitochondrial dynamics may come into play. According to our findings, fusion and then fission allows the cell to take two mitochondria with some damage, and generate two daughters of which one is better and one is worse. In this way the quality of mitochondria can be improved over time. This would explain why in the absence of mitochondrial fusion there should be deterioration of mitochondrial function, even if autophagy is intact.

Turnbull: Then you would predict that if you have problems with these mitochondria in patients' cells, there would be a higher turnover rate than in healthy cells.

Shirihai: Unless in these diseases they have a problem getting into the fusion stage. Therefore they can't improve, and the only selection mechanism is autophagy, which sets the level at whatever it is selecting for.

Turnbull: Then activity would never go to a higher level, which it clearly does.

Schon: It usually only goes to a high level after passing through the germline. Most mutations, with the exception of MELAS, are constant through life. You die with what you are born with.

Turnbull: I was thinking about ageing. It is the same phenomenon in ageing, yet we find cells with remarkably high levels of mutations. I don't understand why these cells aren't cleared, but obviously they are not being cleared in the substantia nigra. It seems to me that perhaps autophagy isn't selecting on this basis. It seems curious that we are not losing this, because even if autophagy was only partially effective, we'd be surprised if it ever got to that level.

Scorrano: Can autophagy itself be controlled by mitochondria? If so, then we could have an impairment of autophagy in cells where mitochondria are dysfunctional, because of mutation in the mitochondria.

Lemasters: If we damage mitochondria we can induce autophagy. I can only speculate whether certain mutations of mtDNA make mitochondria less susceptible to autophagy, possibly because they make fewer free radicals or change the redox state. One thing that impresses me about both inherited diseases and ageing is how long it takes: there is a long latency before the effects are seen. The accumulation of mutations and so on really takes off in the fifth or sixth decade of life. It might be that mitophagy/autophagy is preventing the accumulation of mtDNA mutations, and later in life mitophagy begins to fail for other reasons.

Scorrano: There is an experiment that could test this hypothesis. This is crossing the mito mutator mouse with the ATG5 knockout, to see whether we get a phenotype that arises earlier.

Turnbull: It might be possible to do this in cells by switching on autophagy. Mice treated by rapamycin show decreased mitochondrial activity.

Scorrano: This is always difficult to interpret, because rapamycin has so many effects.

Turnbull: People who are treating Huntington's mice with rapamycin see a decline in mitochondrial enzyme activity.

Rizzuto: You could also do this in cell lines. This could be combined with the analysis of the parameters that could be affected.

Turnbull: We are trying to do this in our hybrid cell lines, treating them with rapamycin. At least then we have a fairly easy signal because heteroplasmy is easy to measure and specific. If you see a change it might be interesting to look at the mechanism.

Nicholls: We've reached the end of this meeting and it would be superfluous for me to give a formal summing-up. It has been an excellent meeting, and thank you all for your participation.

References

Mandal S, Guptan P, Owusu-Ansah E, Banerjee U 2005 Mitochondrial regulation of cell cycle progression during development as revealed by the tenured mutation in Drosophila. Dev Cell 9:843–854

Murray CM, Hutchinson R, Bantick JR et al 2005 Monocarboxylate transporter MCT1 is a target for immunosuppression. Nat Chem Biol 1:371–376

Orrenius S, Gogvadze V, Zhivotovsky B 2007 Mitochondrial oxidative stress: implications for cell death. Annu Rev Pharmacol Toxicol 47:143–183

Singh B, Halestrap AP, Paraskeva C 1997 Butyrate can act as a stimulator of growth or inducer of apoptosis in human colonic epithelial cell lines depending on the presence of alternative energy sources. Carcinogenesis 18:1265–1270

Contributor Index

Non-participating co-authors are indicated by asterisks. Entries in bold indicate papers; other entries refer to discussion contributions.

A

Adam-Vizi, V. 40, 58, 82, 119, 155, 156
*Affourtit, C. **70**
*Aguiari, P. **122**
*Akar, F. **140**
*Aon, M. **140**

B

Beal, M. F. 67, 82, 83, 165, **183**, 192, 193, 194
Bednarczyk, P. 89
Bernardi, P. 37, 38, 39, 43, 44, 58, 86, 120, 133, 152, 153, **157**, 164, 165, 166, 167, 168, 179, 180, 181
Brand, M. D. 38, 40, 66, **70**, 80, 81, 82, 83, 84, 85, 86, 87, 88, 89, 90, 120, 153, 209

C

*Cortassa, S. **140**
*Crichton, P. G. **70**
*Csordás, G. **105**

D

*De Stefani, D. **122**
*Deutsch, M. **21**
*DiMauro, S. **214**
Duchen, M. 42, 68, 103, 119, 120, 138, 155, 192

F

*Forte, M. **157**

G

Giulivi, C. 17, 81, **92**, 100, 101, 102, 103, 104, 135, 177, 213

H

*Haigh, S. E. **21**
Hajnóczky, G. 37, 44, **105**, 118, 119, 120, 121, 136, 137, 138, 178
Halestrap, A. P. 16, 17, 20, 37, 41, 43, 44, 56, 84, 85, 89, 120, 139, 153, 166, 167, 179, 180, 181, 212, 229, 230

J

Jacobs, H. T. 15, 18, 19, 40, 45, 64, 66, 118, 132, 134, 135, 154, 177, 180, 194, 195, 196, 209, 210, 211, 212, 227, 231
*James, D. **170**

K

*Karbowski, M. **4**

L

Larsson, N.-G. 43, 67, 68, 82, 133, 167, 192, 194, **197**, 208, 209, 210, 211, 212, 213, 226
Lemasters, J. J. 16, 18, 36, 41, 43, 44, 56, 83, 88, 89, 103, 132, 154, 164, 165, 167, 177, 210, 228, 232
*Leo, S. **122**
*Lucken-Ardjomande, S. **170**

CONTRIBUTOR INDEX

M

*Marchi, S. **122**
Martinou, J.-C. 17, 37, 40, 58, 67, 89, 134, 164, **170**, 176, 177, 178, 179, 180, 181, 210, 230
*Molina, A. A. J. **21**
*Montessuit, S. **170**

N

*Neutzner A. **4**
Nicholls, D. G. **1**, 14, 16, 19, 36, 37, 38, 39, 41, 44, 55, 56, 57, 58, 63, 67, 80, 83, 84, 85, 86, 87, 88, 89, 90, 103, 131, 132, 133, 135, 137, 138, 152, 153, 155, 164, 165, 166, 167, 180, 181, 193, 194, 196, 208, 209, 210, 211, 212, 228, 230, 231, 232

O

O'Rourke, B. 40, 56, 57, 68, 84, 103, 133, 137, **140**, 152, 153, 154, 155, 156, 180, 211
Orrenius, S. 55, 65, 66, 85, 102, 134, 136, 177, 181, 227

P

Parekh, A. 136, 137
*Parker N. **70**
*Parone, P. A. **170**
*Pinton, P. **122**

R

Reynolds, I. J. 17, 38, 66, 68, 84, 117, 121, 137, 139, 178
Rich, P. R. 42, 43, 102, 154, 228
*Rimessi, A. **122**

Rizzuto, R. 15, 37, 44, 57, 64, 101, 118, **122**, 131, 132, 133, 134, 135, 136, 137, 138, 195, 229, 232
*Romagnoli, A. **122**

S

*Saotome, M. **105**
Schon, E. A. 18, 36, 42, 44, 45, 86, 100, 101, 139, 167, 178, 181, 194, 195, 196, 213, **214**, 226, 227, 228, 229, 230, 231, 232
Scorrano, L. 19, 20, 38, 44, 45, **47**, 55, 56, 57, 58, 59, 65, 119, 134, 135, 138, 176, 178, 179, 180, 230, 232
Shirihai, O. 15, 18, **21**, 36, 38, 39, 40, 41, 42, 43, 44, 67, 89, 90, 137, 138, 177, 178, 180, 211, 226, 230, 231, 232
Spiegelman, B. M. 16, 19, 20, 55, **60**, 64, 65, 66, 67, 68, 80, 81, 87, 88, 101, 102, 104, 132, 192, 193, 194, 208, 209, 212, 226, 227

T

*Terradillos, O. **170**
*Terzioglu, M. **197**
Turnbull, D. M. 42, 65, 90, 193, 194, 195, 211, 213, 227, 228, 229, 231, 232
*Twig, G. **21**

W

*Weaver, D. **105**
*Wikstrom, J. D. **21**

Y

*Yi, M. **105**
Youle, R. J. **4**, 14, 15, 16, 17, 18, 19, 37, 57, 58, 100, 138, 165, 177, 178, 179, 181, 231

Z

*Zecchini, E. **122**

Subject Index

A

α-ketoglutarate dehydrogenase 185
α-synuclein 186
AAA proteases 4, 5
ABAD (amyloid β binding alcohol dehydrogenase) 185
acyl-CoA dehydrogenases 90
adaptive thermogenesis 73
adenine nucleotide translocator (ANT) 85–86, 89, 157–159, 166–168
aequorin 123, 128
ageing
    apoptosis 197, 205–206, 208–209
    caloric restriction 198–199
    disposable soma theory 198
    genetics of mtDNA 200–203
    mitochondrial dysfunction 197–213
    neurodegeneration 183–184, 193
    oxidative stress 197, 204–205, 208–209, 212–213
    respiratory chain 197, 199–201, 204, 209–212
    single gene mutants 199
    somatic mtDNA mutations 202–203
AIF 2, 55, 124–125, 170
Akt 68
alamethicin 168
alkylsulfonates 71
Alpers syndrome 222
ALS *see* amyotrophic lateral sclerosis
Alzheimer's disease 183, 184–185, 192, 194
AMP 85, 87
amyloid β 184–185
amyotrophic lateral sclerosis (ALS) 183, 187–188, 192
ANT *see* adenine nucleotide translocator
antioxidants 65–66, 83
Apaf-1 93
apoptosis
    endoplasmic reticulum cross-talk 124–125, 126, 134, 136
    mitochondria-shaping proteins 47–59
    mitochondrial dysfunction 197, 205–206, 208–209
    neurodegeneration 184, 187
    outer mitochondrial membrane 170–171, 172, 179
    permeability transition pore 160, 165–166
    respiratory chain 229–230
APP (amyloid precursor protein) 184–185
arrhythmias 148
astrocytes 38
Atkins diet 228–229
ATP synthase 1, 39
    ageing 199
    Ca^{2+} signalling 118
    respiratory chain 217
ATPases 56, 95
atractyloside 56, 84–86, 158–159, 166–167
atrogenes 65
atrogin 62
autophagocytosis 43–44
autophagy 19, 43, 231–232
8-azido-ATP 84

B

β-cells
    PGC1 coactivators 67
    photoactivatible-GFP 27, 38–39
    uncoupling proteins 70, 75–77, 89–90
β3 adrenergic agonists 193
Baf 177, 179
Bak 170–171, 179–180
Bax 16, 164, 170–175, 176–180
Bcl-2
    endoplasmic reticulum cross-talk 135

SUBJECT INDEX

mitochondria-shaping proteins 55, 58
outer mitochondrial membrane
    170–171
permeability transition pore 164
benzodiazepine receptor (BzR) ligands 142, 144, 152, 157, 159–160
BH3-only proteins 171–172
Bid 50–52
biogenesis
    neurodegeneration 183
    outer mitochondrial membrane 181
    PGC1 coactivators 60, 61, 68
    PGC1α 183
bongkrekic acid 56, 84–85, 158–159, 166
BzR *see* benzodiazepine receptor

C

Ca^{2+} 2
    ageing 212
    biological significance 113–114
    cardiac function 141–142, 149, 154–155
    control mechanisms 109–114
    cytoskeletal support 108–109
    distribution 106–109
    endoplasmic reticulum–mitochondrial coupling 112, 120
    inhibition 118
    intermitochondrial interactions 112–113
    interorganellar interactions 112–113
    membrane potential 118
    mitochondria-shaping proteins 47
    mitochondrial motions 109, 110–111
    morphology 106–108
    motility 105–121
    nitric oxide 93, 94–95
    permeability transition pore 158, 160–161, 164–165, 167–168
    secondary messengers 109–111
    uncoupling proteins 87–88
    wave propagation 33, 124
    see also endoplasmic reticulum cross-talk
CAG trinucleotide 188, 189
calcineurin 111, 160, 165
calmodulin 87–88, 111, 149
caloric restriction (CR) 198–199
calsequestrin 133
carbachol 110
cardiac function
    cell function 149–150
    cell stress responses 141–142

coupled oscillator networks 142–146, 148–150, 152
depolarization 140, 142–143, 148, 152–156
inner membrane anion channel 141–142, 144, 146, 150, 153–154
ion channels 140–156
ischaemia and reperfusion 146–148, 154
long range temporal correlations 145–146
membrane potential 140, 141, 142–149, 152
mitochondrial criticality 142–145, 146–148
permeability transition pore 141–142, 153–154
reactive oxygen species 140–141, 142–151, 152–156
selectivity of ion channels 150
cardiolipin 86, 177, 178, 181
caspases 179, 186
Charcot-Marie-Tooth IIa 48, 224
chloroplasts 37
cholesterol 176–177
Cl^- 152–153
coenzyme Q (CoQ) 216–219
collagen deficiency 164–165
collateral damage 214, 223–224
complex I 186, 189, 211
complex II 211
contact sites 161
CoQ *see* coenzyme Q
COS-7 cells 29, 33, 39
coupled oscillator networks 142–146, 148–150, 152
COX *see* cytochrome c oxidase
CR *see* caloric restriction
cristae remodelling 48–54, 55–56
criticality 142–145, 146–148
cross-talk *see* endoplasmic reticulum cross-talk
cyclins 15
cyclophilin D (CyP-D) 157–158, 160–161, 164–166
cyclosporin A
    cardiac function 154
    mitochondria-shaping proteins 56–58
    permeability transition pore 160–161, 164, 168
CyP-D *see* cyclophilin D
cytochrome b 184
cytochrome c 2

cardiac function 155
mitochondria-shaping proteins 49–52, 55
neurodegeneration 186
nitric oxide 93
outer mitochondrial membrane 170, 172–175, 178–181
permeability transition pore 161
respiratory chain 219
cytochrome c oxidase (COX)
ageing 211
endoplasmic reticulum cross-talk 139
neurodegeneration 184, 185
nitric oxide 96–98, 102–103
respiratory chain 215–218, 220, 221, 227, 229
cytoskeletal support 108–109

D

deletion disorders 195–196, 221
depolarization
cardiac function 140, 142–143, 148, 152–156
endoplasmic reticulum cross-talk 138
permeability transition pore 164, 166
photoactivatible-GFP 34, 38–44
diabetes 61, 66–68, 77
disposable soma theory 198
DJ1 186–187
DNP 88
dominant optic atrophy 48
Drp1 see dynamin-related protein 1
dynamic networks 34
dynamin-related protein 1 (Drp1)
endoplasmic reticulum cross-talk 126
mitochondria-shaping proteins 47, 57
outer mitochondrial membrane 171, 173, 178–179, 181
photoactivatible-GFP 22
proteasome 12, 14–16
dynamins 51, 57
dyneins 108–109, 111

E

E-myristoylation 95
E3 ligases 4, 9, 11–17, 62
ecstasy (MDMA) 73
electron transport chain (ETC) 128, 144
encephalomyopathies 219–222
endonuclease G 170

endoplasmic reticulum cross-talk 2, 122–139
3D structures 126
apoptosis 124–125, 126, 134, 136
Ca^{2+} homeostasis 122, 123, 125–129, 131–139
concepts in Ca^{2+} uptake 122–125
inositol-1,4,5-triphosphate 122–123, 125, 127–129, 135–136
membrane potential 122–123, 132
protein kinase C 126–127, 133–134
protein–protein interactions 128–129
systolic signalling pathways 126–128
voltage-dependent anion channels 128–129, 132, 135–137
endoplasmic reticulum-associated degradation (ERAD) 4, 5–6, 9, 12, 15–16
endoplasmic reticulum–mitochondrial coupling 112, 120
endothelial nitric oxide synthase (eNOS) 94–95
endothelin B1 177
eNOS see endothelial nitric oxide synthase
ERAD see endoplasmic reticulum-associated degradation
ERRα 60
ETC see electron transport chain

F

fatty acids
oxidation 90
uncoupling proteins 70, 75, 80–82, 86
FCCP 100–101, 118
fibre-type switching 61–62, 68
fibroblasts 42
Fis1 12, 22, 47, 57–59
fission
endoplasmic reticulum cross-talk 132–133, 138
fragmentation 32, 34, 57–58, 171, 178–180
mitochondria-shaping proteins 48, 50, 57–59
outer mitochondrial membrane 12, 15, 19–20, 171, 172–174, 178–180
photoactivatible-GFP 29–32, 33, 41–42, 44–45
respiratory chain 223–224, 230–231
flickering membrane potentials 34, 38–40

SUBJECT INDEX

focal respiratory chain dysfunction 204
fragmentation
   mitochondria-shaping proteins 57–58
   outer mitochondrial membrane 171, 178–180
   photoactivatible-GFP 32, 34
Friedrich's ataxia 190, 213
fusion
   endoplasmic reticulum cross-talk 132–133, 137–139
   mitochondria-shaping proteins 48, 50
   outer mitochondrial membrane 171
   photoactivatible-GFP 24–26, 38–45
   respiratory chain 223–224, 231
Fzo1 5–9, 12–13, 14, 17

G

GABA receptors 108–109
GABP 66
genipin 76–77
GFP *see* green fluorescent protein; photoactivatible-GFP
GLP1 67
glucose tolerance 65
glucose-stimulated insulin secretion (GSIS) 70, 75–77, 89–90
GLUT2 76
Glut4 68
glycolysis 229–230
GRACILE syndrome 220
green fluorescent protein (GFP) 126, 128
   see also photoactivatible-GFP
Gresham's law 226
GRIF1 108–109
GRP1 67
GSIS *see* glucose-stimulated insulin secretion

H

heart failure 60–61
heteroplasmic mutations 218
HNE *see* hydroxynonenal
Holliday junction resolvase 226
HtrA2/Omi 170, 174
humanin 178
Huntington's disease 67, 165, 183, 188–190, 192–193
hydrogen peroxide
   endoplasmic reticulum cross-talk 135
   permeability transition pore 165

PGC1 coactivators 60, 62, 66
respiratory chain 228
8-hydroxyguanosine 212
hydroxynonenal (HNE) 72–73, 74
hyperglycaemia 58, 76
hyperpolarization 39
hypoxia 149

I

IMM *see* inner mitochondrial membrane
IMS *see* intermembrane space
inducible nitric oxide synthase (iNOS) 94, 101
inner membrane anion channel (IMAC) 141–142, 144, 146, 150, 153–154
inner mitochondrial membrane (IMM)
   endoplasmic reticulum cross-talk 128
   mitochondria-shaping proteins 55–56
   permeability transition pore 161–162
   photoactivatible-GFP 37–38
   proteasome 6
   respiratory chain 214, 217
iNOS *see* inducible nitric oxide synthase
inositol triphosphate (IP3)
   calcium signalling 110, 120
   endoplasmic reticulum cross-talk 122–123, 125, 127–129, 135–136
   nitric oxide 94, 103
INS-1 cells 29, 33
insulin
   resistance 68
   secretion 70, 75–77, 89–90
intermediate filaments 119
intermembrane space (IMS) 214, 218
intermitochondrial interactions 112–113
interorganellar interactions 112–113
IP3 *see* inositol triphosphate
ischaemia–reperfusion 82, 146–148, 154

J

JC1 22, 36, 42

K

K^+ 2
   cardiac function 141–142, 148, 153, 155
   uncoupling proteins 75–77, 84, 88
K562 cells 29
kainic acid 63–64, 189
Kearns-Sayre syndrome (KSS) 195–196, 221

ketogenic diets 228–229
KIF5A 224
kinesins 108–109, 111
Krebs cycle 105–106
KSS *see* Kearns-Sayre syndrome

L

lactate 210–211
LC3 43
Leber hereditary optic neuropathy (LHON) 219
Leigh syndrome 219, 220
LHON *see* Leber hereditary optic neuropathy
lipid peroxidation 80–82
liposomes 172, 176–177, 180
long range temporal correlations 145–146
luciferase 124
lumenal continuity 24–26, 29, 32–33

M

MARCH5 12
mastoparan 168
maternally-inherited Leigh syndrome (MILS) 219
MCU *see* mitochondrial calcium uniporter
MDM30 13
MDMA (ecstasy) 73
MEFs *see* mouse embryonic fibroblasts
MELAS 221, 228–229, 232
membrane potential
  Ca^{2+} signalling 118
  cardiac function 140, 141, 142–149, 152
  endoplasmic reticulum cross-talk 122–123, 132
  photoactivatible-GFP 22, 26–34, 36, 38
Mendelian-inherited errors 221–222
3-mercaptopropionic acid (3-MPA) 211
MERRF 221, 229
metabolic memory 124
Mfn *see* mitofusins
MG132 5
Mgm1 57
microfilaments 119
MILS *see* maternally-inherited Leigh syndrome
MIM *see* inner mitochondrial membrane
miro 108–109, 118–119
mitochondria-shaping proteins
  apoptosis 47–59
  cristae remodelling 48–54, 55–56

cytochrome c 49–52, 55
fission 48, 50, 57–59
fusion 48, 50
Opa1 47, 48–59
Parl 47, 52–54, 58
permeability transition pore 55–57
reactive oxygen species 58–59
mitochondrial
  criticality 142–145, 146–148
  dynamics 21–22
  dysfunction 197–213
  web 22, 33
mitochondrial calcium uniporter (MCU) 122–123, 125, 128, 131–133, 136
mitochondrial nitric oxide synthase (mtNOS) 92, 94–96, 98
mitofusins (Mfn) 15, 17, 22, 47, 48–50
mitosis 15, 218
MitoTrackers 37, 139
MLASA 223
MNGIE 222
MOM *see* outer mitochondrial membrane
mosaic respiratory chain deficiency 197, 201
motility
  biological significance 113–114
  Ca^{2+} signalling 105–121
  control mechanisms 109–114
  cytoskeletal support 108–109
  distribution 106–109
  endoplasmic reticulum–mitochondrial coupling 112, 120
  inhibition 118
  intermitochondrial interactions 112–113
  interorganellar interactions 112–113
  mitochondrial motions 109, 110–111
  morphology 106–108
  second messengers 109–111
mouse embryonic fibroblasts (MEFs) 49–54, 58
3-MPA *see* 3-mercaptopropionic acid
MPTP 62, 63–64, 66, 189
MPV17 gene 222–223
mtNOS *see* mitochondrial nitric oxide synthase
Müller's ratchet 42, 202, 226
MURF 62
muscle tissue 44–45
muscular atrophy 62, 65
mutation events 2–3
myosin motors 108, 111, 118–119
myxothiazol 156

SUBJECT INDEX

N

Na$^+$ ATP channels 88
Na$^+$/H$^+$ exchangers 133
NADH/NADP transhydrogenase (NNT) 90
NARP *see* neuropathy, ataxia and retinitis pigmentosa
neurodegeneration 2, 183–196
    ageing 183–184, 193
    Alzheimer's disease 183, 184–185, 192, 194
    amyotrophic lateral sclerosis 183, 187–188, 192
    Huntington's disease 183, 188–190, 192–193
    Kearns-Sayre syndrome 195–196
    Parkinson's disease 183–184, 186–187, 195
    PGC1 coactivators 60–61, 62
neuronal nitric oxide synthase (nNOS) 94–96, 100–102
neuropathy, ataxia and retinitis pigmentosa (NARP) 219
nigericin 84
nitric oxide (NO) 2, 92–104
    cytochrome c oxidase 96–98, 102–103
    dynamic control 93–96
    inhibition 97, 103–104
    phosphorylation 95–96
    physiological relevance 98
    production and regulation 92–93
    respiratory chain 229
nNOS *see* neuronal nitric oxide synthase
NNT *see* NADH/NADP transhydrogenase
NO *see* nitric oxide
nocodazole 119–120
non-shivering thermogenesis 73
nuclear receptor family (NRF) 60, 66

O

obesity 88
OIP106 108–109
oligomycin 41–44, 118, 119
OMM *see* outer mitochondrial membrane
Optic Atrophy 1 (Opa1) 19, 47, 48–59
outer mitochondrial membrane (OMM)
    apoptosis 170–171, 172, 179–180
    Bax activation 170–175, 176–180
    BH3-only proteins 171–172
    Ca^{2+} signalling 108–109, 112
    cellular distribution 11
    cytochrome c 170, 172–175, 178–181
    domain organization 10
    endoplasmic reticulum cross-talk 128
    ERAD pathway 4, 5–6, 9, 12, 15–16
    fission 12, 15, 19–20, 171, 172–174, 178–180
    Fzo1 5–9, 12–13, 14, 17
    HtrA2/Omi 170, 174
    liposomes 172, 176–177, 180
    permeability transition pore 161–162, 172, 179
    permeabilization 170–182
    photoactivatible-GFP 37
    protein degradation 4–20
    respiratory chain 214
    RING domain proteins 9–13
    Smac/DIABLO 170, 173–174, 178
    ubiquitin 4, 6–9, 11–19
oxidative phosphorylation
    ageing 209, 212–213
    mitochondria-shaping proteins 48
    PGC1 coactivators 61, 68
    respiratory chain 215–218, 231
oxidative stress 2
    ageing 197, 204–205, 208–209, 212–213
    Ca^{2+} signalling 120
    cardiac function 153
    endoplasmic reticulum cross-talk 134
    neurodegeneration 184–185, 187
    permeability transition pore 164, 167–168
    uncoupling proteins 70, 74–75, 81, 87
OXPHOS *see* oxidative phosphorylation

P

p53 227
p66 133–134
PA-GFP *see* photoactivatible-GFP
paraplegin 54
paraquat 62, 66
parkin 11–13, 186–187
Parkinsonism 60
Parkinson's disease 183–184, 186–187, 195, 204
Parl 47, 52–54, 58
PBR *see* peripheral benzodiazepine receptor
PDZ domain 96, 100
Pearson syndrome 221
PEO *see* progressive external ophthalmoplegia

peripheral benzodiazepine receptor (PBR) 157, 159–160
permeability transition pore (PTP) 55–57, 157–169
   adenine nucleotide translocator 157–159, 166–168
   Bax activation 164, 172, 179
   cardiac function 141–142, 153–154
   contact sites 161
   cyclophilin D 157–158, 160–161, 164–166
   depolarization 164, 166
   formation/structure 161–162, 164, 167–168
   peripheral benzodiazepine receptor 157, 159–160
   voltage-dependent anion channels 157, 159
permeabilization 170–182
peroxisome autophagy 19, 43
peroxynitrite 93, 101, 104
PGC1 coactivators
   antioxidants 65–66
   biogenesis 60, 61, 68
   disease 61–62
   fibre-type switching 61–62, 68
   neuronal effects 64–65
   transcriptional control 60–69
PGC1α 2
   biogenesis 183
   endoplasmic reticulum cross-talk 126–128
   neurodegeneration 183, 188–190, 192–194
   respiratory chain 227
   uncoupling proteins 87–88
phenylarsine oxide 167
photoactivatible-GFP (PA-GFP) 21–46
   depolarization 34, 38–44
   dynamic networks 34
   experimental approach 23–24
   fission 29–32, 33, 41–42, 44–45
   flickering membrane potentials 34, 38–40
   fragmentation 32, 34
   fusion 24–26, 38–45
   individual mitochodrion 22, 23–29, 45
   inner membrane proteins 26
   large, non-flat cells 23
   lumenal continuity 24–26, 29, 32–33
   membrane potential 22, 26–34, 36, 38
   mitochondrial dynamics 21–22
   mitochondrial web 22, 33
   morphological/physiological discrepancies 32–33
   non-adherent cells 23–24
   subcellular heterogeneity 27–29, 34
   ubiquitin 40
   z stacking 37, 41
PIN1 *see* prolyl isomerase
PINK1 186–187
PMF *see* protonmotive force
Pol-g mutations 2
PolG 209, 212, 222
power spectral analysis (PSA) 146, 147
Ppif gene 157, 160–161
PRC (PGC1-related coactivator) 64
progressive external ophthalmoplegia (PEO) 221–222
prohibitin 18
prolyl isomerase (PIN1) 184–185
protein degradation 4–20
protein kinases 95, 126–127, 133–134
proteinase K 101
proton conductance 71–73, 80, 89
protonmotive force (PMF) 74, 76, 80, 83–84, 86–87
PSA *see* power spectral analysis
PTP *see* permeability transition pore

R

ragged-red fibres (RRF) 221, 226–227
rapamycin 232
RDA *see* relative dispersion analysis
reactive nitrogen and oxygen species (RNOS) 92–93, 101, 104
reactive oxygen species (ROS)
   ageing 197, 204–205, 208–209, 212–213
   cardiac function 140–141, 142–151, 152–156
   mitochondria-shaping proteins 58–59
   neurodegeneration 183–184, 189, 194
   nitric oxide 92–93
   outer mitochondrial membrane 17
   PGC1 coactivators 60, 62, 64, 66
   photoactivatible-GFP 33, 40
   uncoupling proteins 70, 74–75, 80–90
relative dispersion analysis (RDA) 146, 147
respiratory chain
   ageing 197, 199–201, 204, 209–212
   assembly mutations 214, 220
   autophagy 231–232

collateral damage 214, 223–224
deletion disorders 221
diseases 214, 218–224
indirect mutations 220–223
Mendelian-inherited errors 221–222
mitochondrial genome 214–215
mutations 214–233
nDNA mutations 214, 223–224
oxidative phosphorylation 215–218
subunit mutations 214, 218–219
resveratrol 67
reverse electron transport 80
rhod-2 123
RING domain proteins 4, 9–13
RNOS *see* reactive nitrogen and oxygen species
rotenone 186
RRF *see* ragged-red fibres
Rsp5 13

S

SCO1/2 220, 227–228
SERCA2 133, 135
signal transduction 2
single gene mutants 199
Smac/DIABLO 55, 170, 173–174, 178
SOD *see* superoxide dismutases
somatic mtDNA mutations 202–203
SPG7 224
staurosporin 187
subcellular heterogeneity 27–29, 34
superoxide
  cardiac function 154–155
  nitric oxide 93, 98, 102–104
  uncoupling proteins 72, 76, 85
superoxide dismutases (SOD)
  ageing 205, 209
  neurodegeneration 187–188, 189, 192
  nitric oxide 101–102
  PGC1 coactivators 60, 62, 66
  uncoupling proteins 76
SURF1 220
systolic signalling pathways 126–128

T

tachycardia 148
tBid 171–172

thermogenesis 70, 73
threshold effect 218
TMRE 23, 25, 26–30, 36, 40–42
TMRM 22, 42, 132
transcriptional control 60–69
Troyer syndrome 224
trypsin 100
type 2 diabetes 61, 66–68, 77

U

ubiquinone 81
ubiquitin 4, 6–9, 11–19, 40
uncoupling proteins (UCPs) 1, 70–91
  activation 72–73
  catalysis 71–72
  endoplasmic reticulum cross-talk 128
  fatty acids 70, 75, 80–82, 86, 90
  insulin secretion 70, 75–77, 89–90
  PGC1 coactivators 61, 64
  physiological functions 70, 73–77
  proton conductance 71–73, 80, 89
  proton cycling 84
  protonmotive force 74, 76, 80, 83–84, 86–87
  reactive oxygen species 70, 74–75, 80–90
  thermogenesis 70, 73
  tissue specificity 82

V

vasopressin 118, 120
voltage-dependent anion channels (VDAC)
  calcium signalling 108
  endoplasmic reticulum cross-talk 128–129, 132, 135–137
  permeability transition pore 157, 159

W

Warburg effect 227–228

Y

yellow fluorescent protein (YFP) 11, 49

Z

z stacking 37, 41
Zn^+ 111